CLUSTERING AND ROUTING ALGORITHMS FOR WIRELESS SENSOR NETWORKS

Energy Efficient Approaches

CLUSTERING AND ROUTING ALGORITHMS FOR WIRELESS SENSOR NETWORKS

Energy Efficient Approaches

Pratyay Kuila
Prasanta K. Jana

CRC Press
Taylor & Francis Group
Boca Raton London New York

CRC Press is an imprint of the
Taylor & Francis Group, an **informa** business
A CHAPMAN & HALL BOOK

CRC Press
Taylor & Francis Group
6000 Broken Sound Parkway NW, Suite 300
Boca Raton, FL 33487-2742

Version Date: 20170823

International Standard Book Number-13: 978-1-4987-5382-1 (Hardback)

Visit the Taylor & Francis Web site at
http://www.taylorandfrancis.com

and the CRC Press Web site at
http://www.crcpress.com

Dedicated to my parents.

Pratyay Kuila

Dedicated to my wife, Mrs.Tanusree Jana, and my daughter, Ms.Purbasha Jana.

Prasanta K. Jana

Contents

Preface

The rapid advances in wireless communication and networking, coupled with miniaturization and sensing technology, have made wireless sensor network (WSN) a powerful technology. Over the past few years, the world has witnessed potential applications of WSNs in environment monitoring, surveillance systems, military, health care, disaster warning systems, and home applications. However, in such applications, the major concern is how to judiciously use the energy of the sensor nodes. This is because the sensor nodes have a limited power source. Moreover, in many applications such as battlefields, forests, etc., WSNs are deployed in harsh environments with limited access to human beings. Therefore, it is almost impossible to replace the sensor nodes or replenish their battery when their energy gets completely exhausted. Therefore, efficient usage of energy of the sensors has been the prime focus in the applications of WSNs. Various issues have been studied for reducing energy consumption of the sensor nodes to extend the lifetime of WSNs. Among them, energy-efficient clustering and routing are the two most important topics and have been studied extensively.

The purpose of this book is to provide a comprehensive and systematic introduction to WSNs, their applications, and various effective solutions towards clustering and routing problems in WSNs. The book is written with suitable numerical and figurative illustrative examples and the pseudo code of the algorithms, with their performance evaluation and analysis. Distinguishing it from other books, this book focuses on a survey and details of many energy-efficient clustering and routing algorithms with a great emphasis on the evolutionary approaches that are not considered in other books.

Who Can Use This Book

Graduate students, the researchers, and the developers interested in the field of wireless networks are the main target readers. Specifically, this book will be highly beneficial to all the researchers who wish to pursue research in the algorithm design for clustering and routing algorithms in the field of wireless sensor networks. The book will help learners easily understand the basics of wireless sensor networks, and energy saving schemes for clustering and routing algorithms. They will know how to develop algorithms, and how to apply evolutionary algorithms in solving problems in other areas, too. If used in an academic setting, the material in this book is suitable for a one-semester graduate level course. Although the book is self-contained, a few chapters require some rudimentary knowledge of advanced data structures, advanced algorithm design, soft computing approaches, and distributed algorithms.

About the Authors

Pratyay Kuila received his B.Tech. degree in 2008 and M.Tech. degree in 2011, both in Computer Science and Engineering, from West Bengal University of Technology, Kolkata. He has received his Ph.D. degree in 2014 from the Indian Institute of Technology (ISM), Dhanbad. He was an assistant professor in the School of Computer Engineering, KIIT University, Bhubaneswar, India. Currently, he is an assistant professor in the Department of Computer Science & Engineering, National Institute of Technology, Sikkim, India. He has acted as a referee in many reputable International Journals, including *Ad Hoc Networks*, *Telecommunication Systems*, etc. He has contributed 19 research papers in the field of wireless sensor networks. His main research interest is to develop energy efficient clustering and routing algorithms for wireless sensor networks.

Prasanta K. Jana received his M.Tech. degree in Computer Science from the University of Calcutta in 1988, and his Ph.D. from Jadavpur University in 2000. Currently he is a professor in the department of Computer Science and Engineering, Indian Institute of Technology (ISM), Dhanbad, India. He has about 26 years of teaching and research experience in the field of wireless sensor networks, parallel and distributed computing, algorithm design, and data clustering. He has contributed 160 research publications, co-authored four books, two book chapters and edited two books. He has also mentored seven Ph.Ds. In recognition of his outstanding research contributions, he was appointed senior member of IEEE (The Institute of Electrical and Electronics Engineers, USA) in 2010 and received a Canara Bank Research Publication Award in 2015. He is on the editorial board of two international journals and acted as referee in many reputable international journals. Dr. Jana has acted as the general chair of the International Conference Recent Advances in Information Technology (RAIT)-2012, Co-chair of the National Conference Recent Advances in Information Technology (RAIT)-2009 and Convener of the workshop WPDC-2008. He has also served as advisory committee and programme committee member of several international conferences. His current research interests include wireless sensor networks, cloud computing, big data, and machine learning. He visited the University of Aizu, Japan, in 2003, Las Vegas, USA in 2008, Imperial College of London, UK in 2011, University of Macau, in 2012, and Hong Kong in 2015, for academic purposes.

Acknowledgments

This book is the result of research carried out for Ph.D thesis work by Dr. Pratyay Kuila under the supervision of Prof. Prasanta K. Jana, Indian Institute of Technology (ISM), Dhanbad. The authors are pleased to thank many individuals, without whose contributions this book would not have been possible.

First and foremost, we must acknowledge and thank The Almighty Lord for giving us enough strength, encouragement, wisdom, and support. We would like to express our sincere gratitude to Prof. D. C. Panigrahi, the director of the Indian Institute of Technology (ISM), Dhanbad, and Prof. R. Venugopal, the dean (Research & Development) of the Indian Institute of Technology (ISM) for their moral support and for providing us all the necessary facilities.We are thankful to all the faculty and staff members of our department for their help and support. We would also like to acknowledge all the research scholars who have helped in coding and implementation of the algorithms presented in this book. We specially acknowledge the contributions of Dr. Md. Azharuddin, Dr. Suneet K. Gupta, and Shri Amar Kaswan for the completion of this book.

We sincerely thank all the reviewers, regional editors, and chief editors of the journals who provided helpful comments and suggestions that significantly improved the quality and organization of the research articles presented in this book. These editors include Dr. Swagatam Das of *Swarm and Evolutionary Computation* (Elsevier), Dr. Ajith Abraham of *Applied Soft Computing* (Elsevier), Dr. Mario Köppen of *Engineering Applications of Artificial Intelligence* (Elsevier), and Dr. M. Malek of *Computers & Electrical Engineering* (Elsevier). We pay our sincere thanks to Mrs. Astha Sharma, senior acquisitions editor and Shikha Garg, editorial assistant, CRC Press-Taylor & Francis Group, Taylor & Francis India Pvt. Ltd., for their cooperation, help, and support.

We are indebted to our family members, without whose love and support this work could not have been completed. We are also indebted to our parents, for everything that they have given us and sacrificed during the preparation of this book. They have stood by us in everything we have done by providing constant support, encouragement, and belief in us to complete this journey.

Pratyay Kuila and Prasanta K. Jana

Organization of the Book

The first chapter presents a brief introduction to WSNs, along with the major constraints, need, and benefits of clustering and routing algorithms. Then we provide a brief survey of several existing energy-efficient clustering and routing algorithms for WSNs in the second chapter. In the third chapter, major applications of the WSNs are described. In the next five subsequent chapters, we present energy-saving clustering and routing algorithms, which are divided into five basic categories, i.e., centralized algorithms, evolutionary approaches, distributed algorithms, fault-tolerant algorithms and algorithms for hot-spot or energy-hole problems. For the centralized approaches, we use greedy methods, approximation algorithms, and several advanced data structures. For the evolutionary-based approaches, we present various clustering and routing algorithms using genetic algorithm (GA), differential evolution (DE) and particle swarm optimization (PSO). The distributed algorithms are presented with a cost-based approach. Some clustering and routing algorithms are also presented to solve fault-tolerant and the hot-spot or energy-hole problems in wireless sensor networks. For the ease of understanding, all the algorithms are described with illustrative examples. The analysis of the experimental parts of the algorithms are nicely described with suitable diagrams. They are also supported by lemmas and theorems with their proofs.

List of Tables

List of Figures

List of Algorithms

Chapter 1
Introduction

Chapter 1

Introduction

1.1 Background of wireless sensor networks

Rapid advances in micro-electro-mechanical-system (MEMS), computer networks, and wireless communication technology have enabled wireless sensor networks (WSNs) as one of the fastest growing technologies. WSNs have attracted enormous attention for their potential applications in diverse areas such as disaster warning systems, environment monitoring, health care, safety, surveillance, intruder detection, and others [1, 2, 3].

The sensor nodes have limited sensing, processing, and communication capabilities. Most importantly, they have a limited and irreplaceable power source. Moreover, in many applications, it is almost impossible to replace the sensor nodes when their energy is completely exhausted. Therefore, energy consumption of the sensor nodes is the most challenging problem for the long-run operation of WSNs [4, 5]. Various issues have been studied for this purpose and include low-power radio communication hardware [6, 7], energy-aware medium access control (MAC) layer protocols [8, 9, 10], and duty-cycling [4, 11, 12, 13]. However, energy efficient clustering and routing are the two most important topics to have been extensively addressed by the research community.

One of the main advantages of WSNs is their ability to operate in harsh or unattended environments, such as in deep forest, under water, etc., where monitoring the network by humans is risky, inefficient, and sometimes infeasible. The sensor nodes are deployed in the target area in either an ad-hoc or pre-planned manner [14]. Ad-hoc deployment of the sensor nodes is useful for harsh environments that are hard to access, and in such areas, sensor nodes may be dropped by a helicopter. However, this approach requires deployment of a large number of sensor nodes to ensure full coverage of the region. The pre-planned or manual deployment is used for easily accessible regions. This kind of deployment brings better network management and saves energy of the sensor nodes. Moreover, it also minimizes the overall network costs as covering an area with the minimum number of sensor nodes can be possible in manual deployment.

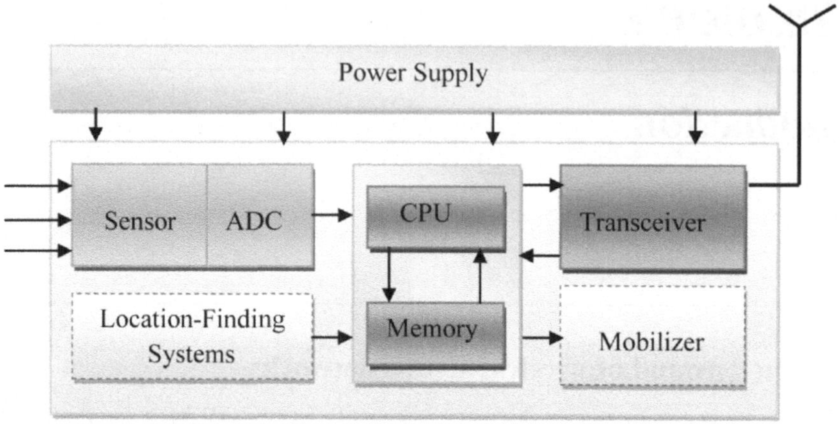

FIGURE 1.1: Basic components of a sensor node

1.2 Structure of sensor nodes

Generally a sensor node consists of four main components: (i) a sensing subsystem including one or more sensors (with associated analog-to-digital converters) for data acquisition; (ii) a processing subsystem including a micro-controller and memory for local data processing; (iii) a radio subsystem or transceiver for wireless data communication; and (iv) a power supply unit as shown in Figure 1.1. Sensor nodes produced by Crossbow are shown in Figure 1.2. In many applications, sensor nodes may also include some additional components, which include a location finding system such as global positioning systems (GPS) to determine their position, a mobilizer to move within the targeted sensing location, and so on.

Sensing unit: The sensing unit is the main component of a wireless sensor node that distinguishes it from any other embedded system with communication capabilities. The sensing unit may generally include several sensor units, which provide information gathering capabilities from the physical world. Each sensor unit is responsible for gathering information of a certain type, such as temperature, humidity, or light, and is usually composed of two subunits: a sensor and an analog-to-digital converter (ADC). The analog signals produced by the sensor based on the observed phenomenon are converted to digital signals by the ADC, and then fed into the processing unit.

Processing unit: The processing unit is the main controller of the sensor node. This unit manages every other components. The processing unit may consist of an on-board memory or may be associated with a small storage unit. This unit manages the procedures that enable the sensor node to perform sensing operations, run associated

(a)

(b)

FIGURE 1.2: Sensor nodes produced by Crossbow: a) Mica2 and b) TelosB

algorithms, and collaborate with the other nodes through wireless communication.

Transceiver unit: Communication between any two sensor nodes is performed by the transceiver units. A transceiver unit implements the necessary procedures to convert bits to be transmitted into radio frequency (RF) waves and recover them at the other end. Essentially, the WSN is connected to the network through this unit.

Power unit: One of the most important components of a sensor node is the power unit. Usually, battery power is used, but other energy sources are also possible. Each component in the wireless sensor node is powered through the power unit and the limited capacity of this unit requires energy-efficient operation for the tasks performed by each component.

Location-finding system: Most of the sensor network applications, sensing tasks, and routing techniques need knowledge of the physical location of a node. Thus, it is common for a sensor node to be equipped with a location-finding system. This system may consist of a GPS module for a high-end sensor node or may be a software module that implements the localization algorithms that provide location information through distributed calculations.

Mobilizer: A mobilizer may sometimes be needed to move sensor nodes when it is necessary to carry out the assigned tasks. Mobility support requires extensive energy resources and should be provided efficiently. The mobilizer can also operate in close interaction with the sensing unit and the processor to control the movements of the sensor node.

Power generator: While battery power is mostly used in sensor nodes, an additional power generator can be used for applications where longer network lifetime is essential. For outdoor applications, solar cells are used to generate power. Similarly, energy scavenging techniques for thermal, kinetic, and vibration energy can also be used [15].

1.3 WSN architecture

A WSN consists of a large number of tiny, low-power, and inexpensive sensor nodes, which are randomly or manually deployed over an unattended target area. The sensor nodes periodically collect local information, process it and finally send it to a remote base station (BS), called a sink either through single-hop or multi-hop communication. The sink is connected to the Internet for the public notice of the phenomena. Depending on the various applications of the sensor network, the sensed data are communicated to the sink in a continuous, event-driven, query-driven or hybrid manner. In continuous data communication, all sensor nodes send the data periodically to the sink. In some applications, a sensor node transmits the sensed data to

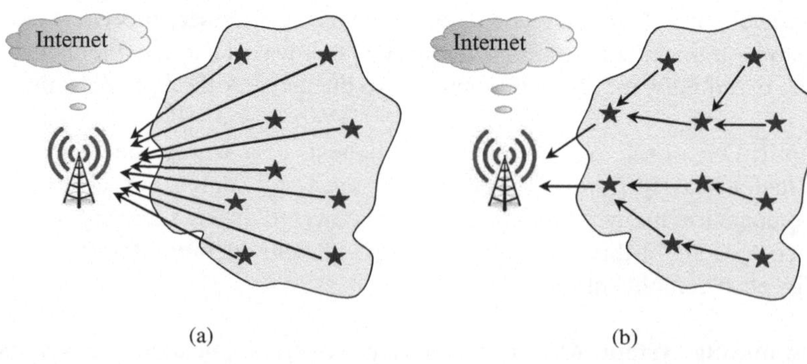

FIGURE 1.3: Flat wireless sensor network architecture: a) single-hop communication and b) multi-hop communication between sensor nodes and sink

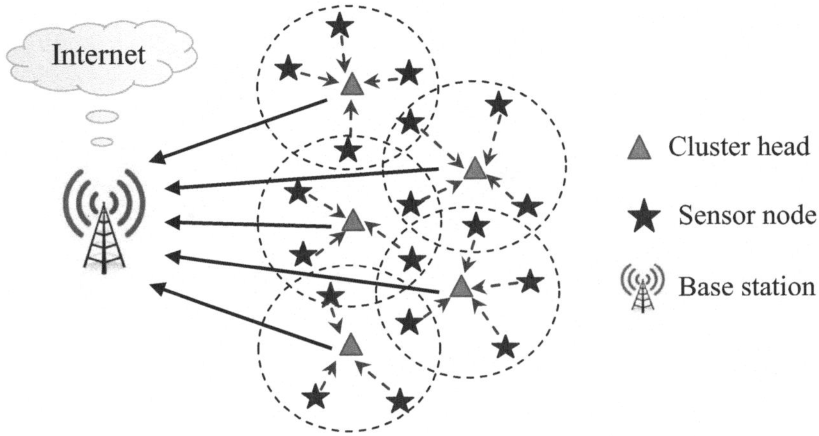

FIGURE 1.4: A cluster-based wireless sensor network architecture with single-hop communication between CHs and base station. Small dashed arrows show the communication between sensor nodes and their corresponding CHs, and large arrows for the same between CHs and CHs or CHs and base station.

the sink only if some event has occurred, i.e., it is event-driven. In the query-driven model, the BS or sink transmits a query to all sensor nodes in the sensing region via flooding or direct communication. Then the sensor nodes that have the data as per the query will respond to the sink. Some applications use a hybrid model, which is a combination of continuous, event-driven, and query-driven data delivery. There are two types of WSN architecture in the literature: 1) flat and 2) hierarchical.

Flat architecture. In a flat sensor network, each sensor node has the same responsibility to efficiently execute the sensing task. The sensor nodes transmit the sensed data to the sink via single-hop or multi-hop communication. The single-hop communication as shown in Figure 1.3 (a) is applicable for a small area sensor network and the sink is located near to the region. Multi-path communication is used for a large area network where each sensor node selects another sensor node as a next-hop relay node to forward the data to the sink. Figure 1.3 (b) shows typical architecture of a flat sensor network with multi-hop communication.

hierarchical architecture. In a hierarchical network, the sensor nodes are divided into several groups called clusters. Each cluster has a leader known as a cluster head (CH). All the sensor nodes sense local data and send it to their corresponding CH. The CHs then aggregate the local data and finally send it to the base station (BS) directly or via other CHs. The functionality of a cluster-based WSN with single-hop and multi-hop communication between CHs and base station can be seen in Figure 1.4 and Figure 1.5, respectively.

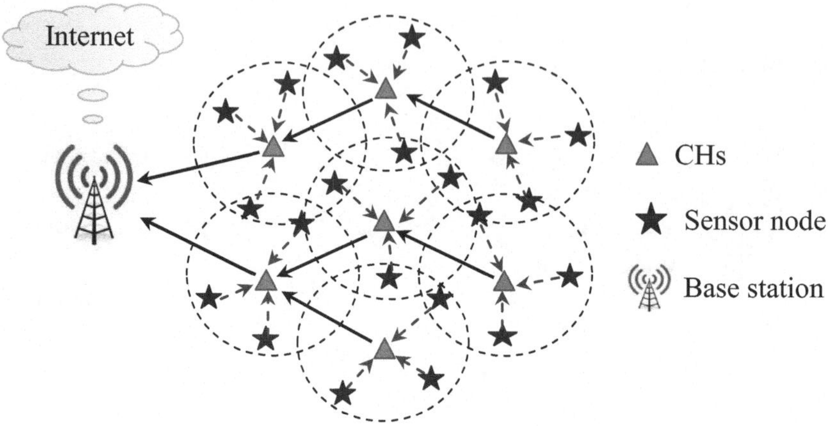

FIGURE 1.5: A cluster-based wireless sensor network architecture with multi-hop communication between CHs and base station

---

1.4 WSNs vs. ad-hoc networks

The main reasons behind resemblance of WSNs to ad-hoc networks are as follows. 1) Both are distributed wireless networks, which indicates there is not a significant network infrastructure in place to build the network. 2) For both of the networks, routing between two nodes may involve the use of intermediate relay nodes, i.e., multi-hop routing. 3) Moreover, both ad-hoc and sensor nodes are operated by small battery. Therefore, there is great anxiety about minimizing the energy consumption of the battery, to extend network life and connectivity. 4) Finally, self-organization is one of the essential requirements of the network because of the distributed nature of both types of networks. Although there are significant resemblances between WSNs and ad-hoc networks, there are also fundamental differences.

- Ad-hoc networks or mobile ad-hoc networks (MANETs) are generally close to humans. Most of the nodes in the ad-hoc network are devices that are made to be operated by humans, e.g., mobile phones, laptops, Personal Digital Assistants (PDAs), etc. On the other hand, sensor nodes are not made to be operated by humans; instead, more focus is given to interaction between sensor nodes and the environment.

- The number of nodes in the WSNs can vary from several hundred to thousands which is larger than the number of nodes in ad-hoc networks. Therefore, scalability of the sensor networks is one of the major issues. There is also a significant variation in node density of sensor networks.

- Sensor nodes are more prone to failure and energy constraints, as they are operated by small batteries. Even so, in many applications, it is almost impossible to replace the sensor nodes when their energy is completely depleted. Therefore, energy consumption for the sensor nodes is one of the most challenging issues for WSNs.

- Sensor nodes may not have any global identifications, so unique addressing is not always feasible in sensor networks.
- Data aggregation is one of the most important features in sensor networks but is not that much important for ad-hoc networks. Generally, the sensor nodes aggregate the local data before sending it to the next node. Here, the main objective is to maximize bandwidth utilization, and reduce energy consumption in data communication.

1.5 Challenges in WSNs

The unique characteristics of sensor networks and their specific applications have a significant impact on facing the various challenging issues in the design of sensor networks. Most of the sensor networks are application-specific, and they have different application requirements. This is one of the main reasons that it is still not possible and feasible to answer all the design challenges in a single network. Instead, only a part of these challenges are considered as per the application requirements in the design of an application specific network. The main challenging aspects of sensor networks are as follows.

- **Limited and irreplaceable energy source:** Reducing energy consumption of the sensor node is one of the most challenging issues. Generally, sensor nodes are operated by small and limited battery power and it is very hard or even impossible to frequently change or recharge the batteries. As an example, the capacity of the Smart Dust mote is 33mAh [16] as powered by a single size 5 hearing aid battery, while MicaZ and Mica2 nodes [17] use two AA batteries with limited capacity 1400 to 3400 mAh. This limited and irreplaceable energy source of the sensor nodes introduces many challenges in the design of hardware, software, and network protocols for WSNs.
- **Scalability:** In many applications, the number of deployed sensor nodes may vary on the order of hundreds or thousands. Hence, whenever designing some new algorithms or protocols for sensor networks, it is quite essential that they will work efficiently and effectively on different network sizes.
- **Fault tolerance:** Sensor nodes are quite prone to failure due to several factors, such as environmental hazards, energy depletion, and device failure. The failure can affect the overall network lifetime and degrades the overall performance of the network. Moreover, failure of the CHs is catastrophic, as it

can limit the accessibility of the sensor nodes under their supervision, and prevent data aggregation and data dissemination. Therefore, in order to keep WSNs operational, the algorithms and protocols for WSNs should cope with the fault-tolerant aspects.

- **Self-organization:** In many applications, sensor nodes are deployed in a harsh environment, and hence, the network is expected to function in an unattended manner or be capable of self organization.
- **Limited processing, storage, and communication:** Sensor nodes have limited processing, storage, and communication capacities. Therefore, they can only perform limited computational functionalities and can communicate within a certain range. These hardware constraints bring in many challenges in software development and network protocol design for WSNs. All algorithms or protocols for sensor networks should consider not only the energy limitations of sensor nodes, but also the processing, storage, and communication limitations.

1.6 System models

1.6.1 Network model

Different network models and designs have been considered for various applications of WSNs. We describe here some of the relevant models used in this book.

Homogeneous and heterogeneous sensor network: Sensor networks can be classified into two types, homogeneous and heterogeneous. In homogeneous networks, all the sensor nodes are identical in terms of energy, processing, and memory capacity. Therefore, in the case of clustering, CHs are selected from the normal sensor nodes. On the other hand, in a heterogeneous sensor network, two or more different types of nodes with different energy levels and functionality are used. Here, comparably less energy-constraint nodes are used as CHs or local sinks; sometimes these are also called gateways. A sensor node can be assigned to any gateway if it is within the communication range of the sensor node. Therefore, there are some pre-specified gateways onto which a particular sensor node can be assigned. Thus, each sensor node has a list of gateways and it can be assigned to only one gateway among them. Similar to LEACH [18, 19], the data-gathering operation is divided into rounds. In each round, all sensor nodes sense local data and send it to their CH. Then CHs perform data aggregation to discard the redundant and uncorrelated data and send the aggregated data to the base station. Between two adjacent rounds, all nodes turn off their radios to save energy. All communication is over wireless link. A wireless link is established between two nodes only if they are within communication range of each other.

Static and mobile sink: Depending on various application scenarios, sinks can be static or mobile. Generally, in a harsh environment, the sink is placed near to the sensing region and the sensor nodes report the sensed event to the static sink through multi-hop or single-hop communication. One of the major problems of static sink scenarios is that the sensor nodes, that are closer to the sink will consume more energy due to relaying the sensed data to the sink, as a result their energy depletes quickly. Subsequently, these sensor nodes die quickly and isolate the sink from the entire network. This problem is commonly known as the hot-spot or sink-hole problem.

To overcome the above problem, another approach is introduced to extend the lifetime of the nodes closer to the sink: is the utilization of a mobile sink. This not only balances energy consumption among the sensor nodes, but also develops connections between the isolated sectors of the network. Another motivation for introducing a mobile sink in WSNs is that some applications explicitly require sink mobility in the targeted field. A mobile sink can collect the data by following different types of mobility patterns, such as random mobility, controlled mobility, or fixed-path mobility. Although the sink mobility improves network lifetime, it invites additional overhead in designing the routing protocol due to sink mobility, for which the topology of a WSN becomes dynamic. The routing algorithms specifically designed for static-sink WSNs cannot be directly applied in a mobile-sink scenario. Another challenges in mobile sink WSNs is the design of strategies to find out the efficient mobility pattern of the sink by taking care of collecting important sensed data in a timely fashion, balancing and reducing energy consumption of the sensor nodes.

Single sink and multiple sink: As static-sink network creates hot-spot or sink-hole problems, another way to solve these problems is to deploy multiple static sinks. The multiple static-sink model reduces the average transmission distance from source to sink and thus reduces energy consumption compared to the single static-sink model. Generally, in the multi-sink scenario each sensor node routes the data to its closest sink. The main challenging issue in the multiple-sink model is the design of a strategy to place the sinks to balance the energy consumption.

Network lifetime: Various definitions of the network life are given in the literature [20, 21], such as the time until the first node dies, the time until the last node dies, or the time until a desired percentage of nodes die. Moreover, in some scenarios [22], network life is considered as the period until the entire region is covered. In [23], the authors have defined the network life in several ways, e.g., N-of-N lifetime, K-of-N lifetime, and m-in-K-of-N lifetime. Where N-of-N lifetime defines the duration until the first gateway dies, K-of-N lifetime presents the duration up to when K gateways out of N are alive, and m-in-K-of-N lifetime presents the duration until all m supporting gateways and overall a minimum of K gateways are alive [23].

1.6.2 Energy model

The radio model for energy used is the same as discussed in [18, 19]. In this model, both the free space and multi-path fading channels are used, depending on the distance between the transmitter and receiver. When the distance is less than a threshold value d_0, then the free space (fs) model is used; otherwise, the multi-path (mp) model is used. Let E_{elec}, ε_{fs}, and ε_{mp} be the energy required by the electronics circuit and by the amplifier in free space and multi-path, respectively. Then the energy required by the radio to transmit an l-bit message over a distance d is given as follows:

$$E_T(l,d) = \begin{cases} lE_{elec} + l\varepsilon_{fs}d^2 & \text{for } d < d_0 \\ lE_{elec} + l\varepsilon_{mp}d^4 & \text{for } d \geq d_0 \end{cases} \tag{1.1}$$

The energy required by the radio to receive an l-bit message is given by

$$E_R(l) = lE_{elec} \tag{1.2}$$

The E_{elec} depends on several factors, such as digital coding, modulation, filtering, and spreading of the signal, whereas the amplifier energy, $\varepsilon_{fs}d^2/\varepsilon_{mp}d^4$, depends on the distance between the transmitter and the receiver and also on the acceptable bit-error rate. It should be noted that this is a simplified model. In general, radio wave propagation is highly variable and difficult to model.

1.6.3 Fault model

There are various reasons for the failure of WSNs. Firstly, sensor nodes may fail due to the depletion of their battery power, malfunctioning of hardware components (such as a processing unit, transceiver, etc.), or damage by an external event. Secondly, the wireless links of the WSNs may fail due to being permanently or temporarily blocked by an obstacle or environmental condition [24]. The link failure causes the network partitions and dynamic changes in network topology. In this book, we consider only permanent failure of the sensor nodes and do not consider the transient and Byzantine failure of the WSNs. We assume that the permanent failure of the sensor nodes due to failure of the hardware components, or complete energy depletion, or permanent link failure. In this book, we consider the reliability of the sensor nodes following the Weibull distribution [25, 26], which is extensively used for fault modeling as it can provide diverse failure patterns over time with its parameters. The probability density function of the Weibull distribution has the following form:

$$f(t) = \frac{\beta}{\eta}\left(\frac{t}{\eta}\right)^{\beta-1} e^{-\left(\frac{t}{\eta}\right)^{\beta}} \tag{1.3}$$

where β and η are the shape and the scale parameters, respectively. This function indicates the likelihood of failure at time t. The reliability function of the Weibull distribution is given by

$$R(t) = e^{-\left(\frac{t}{\eta}\right)^{\beta}} \tag{1.4}$$

The reliability function is the probability that a device is functioning at time t. The failure rate depends on the shape parameter β as follows. When $\beta = 1$, the failure rate is constant; when $\beta > 1$, the failure rate increases over time, and when $\beta < 1$, the failure rate decreases over time, i.e., the component strengthens or hardens over time. Note that $\beta > 1$ for wireless sensor networks.

1.7 Clustering and routing algorithms

In a cluster-based architecture, the sensor nodes are divided into several groups called cluster. Each cluster has a leader known as the cluster head (CH). All the sensor nodes sense local data and send it to their corresponding CH. The CHs then aggregate the local data and finally send it to the base station (BS) directly or via other CHs. The functionality of a cluster-based WSN can be seen in Figure 1.4 and Figure 1.5. A cluster-based WSN has the following advantages [27, 28, 29]:

1. It enables data aggregation at the cluster head to discard the redundant and uncorrelated data. It should be noted that the energy consumed to transfer one bit of data can be used in a high volume of data aggregation. Thus, it reduces energy consumption of the network by preventing it from transmitting high volume of redundant data rather than aggregated data.

2. Routing can be more easily managed because only CHs need to maintain the local route setup of other CHs, and thus require little routing information. This, in turn, improves the scalability of the network significantly.

3. It also conserves communication bandwidth, as the sensor nodes only communicate with their CHs, and thus avoid exchange of redundant messages among themselves.

However, CHs bear some extra workload contributed by their member sensor nodes as they receive the sensed data from their member sensor nodes, aggregate it and communicate it to the BS. Moreover, in many WSNs, the CHs are usually selected among the normal sensor nodes which can die quickly for this extra workload. In this context, many researchers [30, 31, 32, 33, 34, 35, 36, 37, 38, 39] have proposed the use of some special nodes called gateways or relay nodes, which are provisioned with extra energy. These gateways act like cluster heads and are responsible for the same functionality as the CHs.

Unfortunately, the gateways are also battery operated and hence power constrained. The lifetime of the gateways is very crucial for the long-run operation of the network. This is because they manage the network in the form of clusters, and act as an intermediate relay node in the data routing phase to forward the data packets to the base station from the other gateways. Therefore, the death of a few gateways can hamper the whole network activity. It is noteworthy that the transmission energy

(E) which mainly dominates the overall energy consumption is proportional to the distance (d) between transmitter and receiver, i.e., $E \propto d^{\lambda}$, where λ is the path-loss exponent and $2 \leq \lambda \leq 4$ [40]. Therefore, minimization of transmission distance can reduce the energy consumption. However, some applications are very time-critical in nature. Hence, they should satisfy strict delay constraints so that the BS can receive the sensed data within a specified time bound. But the delay is proportional to the number of forwards on the dissemination path between a source and the BS [40]. In order to minimize the delay, it is necessary to minimize the number of forwards, which can be achieved by maximizing the distance between consecutive forwards. Therefore, while designing routing algorithms, we need to incorporate a trade-off between energy consumption and delay, as they pose two conflicting objectives.

Furthermore, the gateways that are closer to the base station or on some energy-efficient path are inevitably used as relay node to forward the packet to the base station, thereby draining their energy very quickly. Therefore, load-balanced clustering is another important issue for WSN clustering. Particularly, this is a pressing issue when the sensor nodes are not distributed uniformly. It is also worth noting that in order to balance the load of the CHs, some sensor nodes are assigned to a CH which may be farther from it. As a result their energy may be drained out due to long haul transmission to the CH and they may quickly die. Therefore, while designing clustering algorithms, one should take care not only the load balancing of the CHs (to save energy of the CHs) but also, energy consumption of the sensor nodes for long run operation of the WSNs.

In the routing phase, CHs may use (i) the single-hop data transmission model (SHDTM) or (ii) the multi-hop data transmission model (MHDTM) to communicate with the base station (BS) [31, 41]. In SHDTM, each CH transmits its data directly to the BS. On the other hand, in MHDTM, the CHs generally use some intermediate CHs that act as relay node(s) to forward the data to the BS [31, 32, 33]. However, the CHs that are closer to the BS or on some energy-efficient path are frequently used as the relay nodes and thus they may drain their energy quickly. Therefore, selection of proper next hop relay nodes is called for in research for designing energy-saving routing in WSNs.

1.7.1 Challenges in WSN clustering

- **Cluster-head selection:** Though clustering is a promising technique to reduce energy consumption, and improve the scalability and the lifetime of WSNs, in cluster-based architecture, CHs bear some extra workload for data receiving, aggregation, and inter cluster communication. Therefore, selection of CHs from the normal sensor nodes is a very crucial issue. Moreover, in the data-routing phase, CHs send the processed data through other CHs in which intermediate CHs act as relay nodes. But this requires the guarantee of connectivity among the CHs along the routing path. It is worth noting that the intermediate non-CH sensor nodes may be used as relay nodes between two unconnected CHs, but this may require special algorithmic design and also generate extra control-message overhead on the network, which may lead to serious energy inefficiency. Nevertheless, there may not always be a sensor node between two unconnected CHs that can act as a relay node. Therefore, proper care should be taken in the selection of CHs, especially whenever CHs are selected from the normal sensor nodes.

- **Cluster formation:** After selection of CHs, it is essential to efficiently assign the normal sensor nodes to their corresponding CHs so that the overall energy consumption can be reduced. Many protocols such as LEACH [18] use the strategy that each non-CH sensor node will join their nearest CH to form a cluster. But this method seems to be inefficient in terms of energy balancing of the network.

- **Load balancing:** If the sensor nodes are not properly assigned to their CHs for cluster formation, then some CHs may be overloaded with a high number of sensor nodes. Such overload may increase latency in communication and degrade the overall performance of the WSN. Therefore, load balancing of the CHs is the most important issue for clustering sensor nodes. Particularly, this is a pressing issue when the sensor nodes are not distributed uniformly. It is also worth noting that in order to balance the load of the CHs, some sensor nodes are assigned to a CH that may be farther from them. As a result, their energy may be drained out due to long-haul transmission to the CH and they may quickly die. Therefore, while designing a clustering algorithm, one should take care not only with the load balancing of the CHs (to save energy of the CHs) but also with energy consumption of the sensor nodes, to increase the network lifetime.

- **Fault tolerant:** Sensor nodes are very prone to failure, which is always harmful for the network. However, failure of the CHs is catastrophic as it can limit the accessibility of the non-CH sensor nodes under their supervision, and prevents data aggregation and data dissemination. Therefore, in order to keep the WSN operational, the clustering and routing algorithms should cope with the fault-tolerant aspects, especially failure of the CHs.

1.7.2 Challenges in WSN routing

- **Proper next-hop selection:** Performance of the network greatly depends on the selection of the next-hop relay node for energy-saving routing. Therefore, proper next-hop relay node selection is an equally important issue.

- **Energy efficiency and energy balancing:** Energy efficiency is a measure of the ratio of total packet delivered at the sink node to the total energy consumed by the sensor nodes (Kbits/J) of the network. The routing algorithm should be energy efficient to prolong the network lifetime. However, in order to be energy efficient, the algorithm is supposed to route the sensed data via other sensor nodes that are closer to the base station or on some energy efficient path. But this arrangement of data routing can lead the sensor nodes of the energy-efficient path to die quickly for they are frequently used as relay nodes. This uneven energy consumption dramatically reduces the network lifetime and decreases the coverage ratio. If the sensor nodes consume their energy more evenly, the connectivity between them and the BS can be maintained for a longer time and thus it prevents the network from being partitioned. Therefore, it is very important to make an appropriate trade-off between energy efficiency and balanced energy consumption.

- **Quality of service:** Some applications are very time-critical in nature. Hence, they should satisfy strict delay constraints so that the BS can receive the sensed data within a specified time bound. Another desired requirement is reliability, by which it is meant that the sensed data should be received by the sink as correctly as possible to ensure accurate decision by the sink.

- **Fault-tolerant and hot-spot problem:** As sensor nodes are very prone to failure, they can affect the overall network lifetime and degrade the overall performance of the network. Therefore, in order to keep a WSN operational, the routing algorithms should cope with the fault-tolerant aspects. Moreover, the sensor nodes closer to the sink are inevitably used as relay nodes to forward the data to the sink, and thereby, drain their energy such that they die quickly. As a result, the particular region becomes vacant and is called a hot-spot area. The routing algorithm should be capable of handle as well this problem.

Chapter 2

WSN Applications

Chapter 2

WSN Applications

Wireless sensor networks (WSNs) have attracted enormous attention for their potential applications in diverse areas such as disaster warning systems, environment monitoring, health care, safety, surveillance, intruder detection, etc. In this section, we briefly describe some important applications of WSNs.

2.1 Forest fire detection

Forest fires are one of the most dangerous natural accidents that occur in almost all countries. Forest fires are a leading disturbance factor in almost all the forest vegetation zones throughout the world. They can cause a huge amount of physical, biological, and environmental losses. Furthermore, forest fires can have hostile social impacts regardless of whether they are caused by natural forces or human activities.

A wireless sensor network (WSN) could be deployed to detect a forest fire in its early stages. A number of sensor nodes would need to be pre-deployed in a forest. The function of this network would be to sense the occurrence of fire at a regular intervals of time and send a message to the sink in the instant a fire is detected. After receiving the warn message, the sink would determine the position of the fire and send an alarm signal to the nearest fire station. Based on the received information, subsequent preventative steps could be initiated to diminish the situation before the fire becomes uncontrollable. The forest fire detection process is graphically presented in Figure 2.1.

2.2 Flood detection

Floods are responsible for the loss of precious lives, and destruction of large amounts of property every year. Several types of sensors are deployed to continuously monitor the rainfall, water level, and weather. These sensors periodically supply information to the local base station and enable the local administration to take necessary actions.

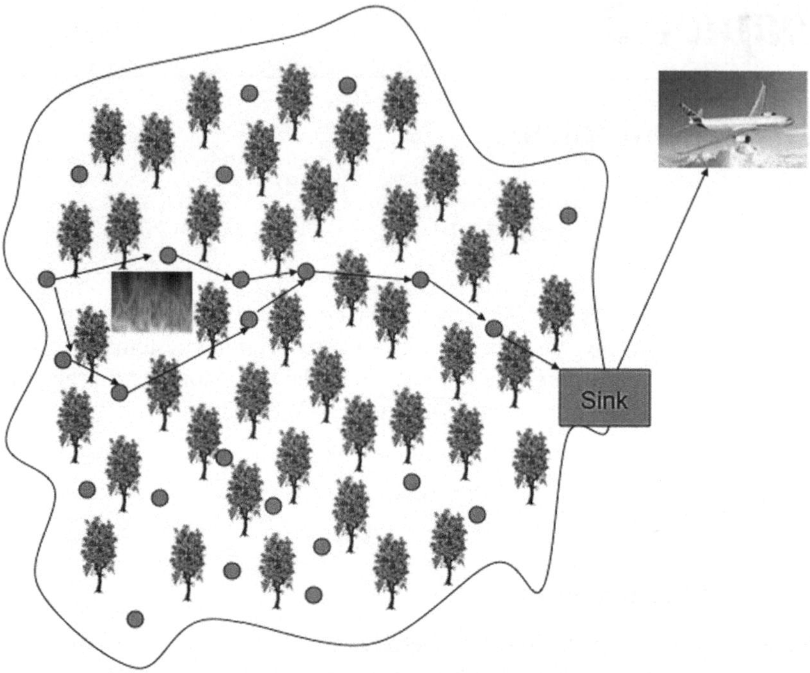

FIGURE 2.1: WSNs for forest fire detection

2.3 Tsunami detection

Tsunamis have become more frequent in recent years due to geographical changes. The intensity of tsunamis and subsequent floods on land can cause enormous financial as well as human loss. A tsunami is usually a consequence of large scale earthquake in deep sea beds. The traditional means of tsunami detection consists of randomly placing underwater sensor nodes in the sea that sense changes in water pressure and store data. With the help of previously stored data, the sensor nodes can predict the possibility of a tsunami.

2.4 Home applications

The frameworks for an intelligent smart home have been proposed recently by the research community. Sensor nodes are used to monitor and report different parameters in a home environment, such as temperature, humidity, and light, as well as controlling different electrical devices for lightning, air conditioning, or heating.

In many cold areas, homes and halls are often heated up by a number of heaters. Controlling the temperature at each separate heater is not always possible or feasible for the living persons in the room. Sometimes, the neighboring rooms have other heating elements that are also influenced by the temperature in the controlled room. WSNs are deployed inside the room for indoor temperature regulation. Sensor nodes are also used in electrical devices, where the sensor can monitor the time and duration of usage of different electrical devices with different power ratings.

2.5 Agricultural applications

Agriculture has played a key role in the development of human civilization. Use of sensor networks in agriculture is headed in a very positive direction. Sensors are used for collecting information about physical and environmental attributes such as weather, crop and soil information. In some regions, the agricultural land is not evenly distributed. Moreover, multiple plants are planted on a single piece of land, which therefore has diverse requirements of crops for different weather and soil conditions. Sensor nodes are deployed in such types of environments to monitor and control different fertilizer and water requirement for different pieces of land.

Figure 2.2 shows an example of the wireless sensor networks in agriculture done in the WaterSense project [42]. The project installs sensor nodes to measure water levels in various drains and the deeper soil, as well as the salinity of water.

FIGURE 2.2: Use of WSNs in agriculture by the WaterSense project

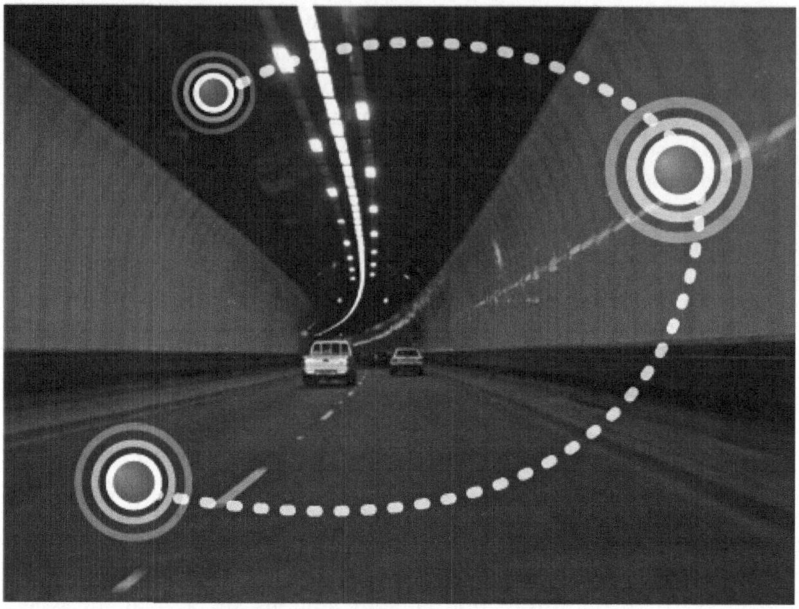

FIGURE 2.3: WSNs in tunnels

2.6 Traffic tracking

Nowadays, traffic jam, and the high number of accidents in urban and metropolitan areas have become more and more stressful and have to dramatic consequences for the economy, human health, and the environment [43]. A WSN-based traffic system is deployed in many applications. One of the main objectives of such a system is to minimize congestion of the whole traffic network and optimize the use of road capacity. This is done through traffic optimization and real-time traffic light control. The deployed sensor nodes transmit information related to accidents in order to reduce congestion and traffic. They can also be used to guide ambulances and fire trucks to reach the exact location of the accident. Sensor networks are also used to create the navigation guidelines for travelers to minimize cost, time, and fuel consumption. In the smart city concept, the sensor networks are used for efficient parking management, that is also a part of traffic minimization.

Wireless sensor networks can also be used to monitor the state of road surfaces. For example, the Massachusetts Institute of Technology (MIT) carried out a research project to detect the number of potholes in a road. Nowadays, many cities are using tunnels due to their geographical regions. WSNs are deployed to monitor the air flow and quality of air, especially for the poisonous gases (CO, CO_2, NO_2) inside tunnels (as shown in Figure 2.3).

2.7 Military applications

The applications of sensor networks in military settings depends not only on the wireless communications technologies, but also on the capabilities of sensors. The sensors can measure various important physical phenomena that are required in military applications including electromagnetic waves, light, pressure and sound, which result from gunfire and explosions. Sensors can detect and possibly measure chemical, biological and explosive vapor, as well as the presence of people or objects. The sensor capabilities are used as one of the key determinants of the type of military application of WSNs [44]. In protected military sites or buildings, unattended acoustic and seismic sensors are deployed to monitor the specific points by the detection of individual enemy soldier's movements.

2.8 Industrial applications

In industry, a sensor can predict the failure of a bearing or of machines. Thus, the mechanics schedule an overnight maintenance and prevent an expensive unplanned shutdown. In some applications, the sensor nodes are deployed to sense some predefined target points like gas cylinders. The sensor nodes can detect and inform the leakage of poisonous gases, give warning before a fire, etc. In a construction site, built-in stress sensors report the structural integrity of buildings, bridges, roadways, and other man-made structures.

2.9 Sensors in education

Sensor technology provides teachers with an exciting alternative to the time-consuming task of manually logging, and observing science experiments. Instead of using stopwatches, thermometers, and barometers, students are using sensors and powerful software to collect and analyze data as they are carrying out their experiments. This provides more accurate readings than the old manual methods, and data is collected in a more exciting way.

2.10 Health care

The medical applications of wireless sensor networks aim to improve existing health care, and monitoring services especially for the elderly, children, and the chronically ill. There are several benefits achieved with these systems. To begin with, remote monitoring capability is the main benefit of pervasive health-care systems. With remote monitoring, the identification of emergency conditions for patients will become easy, and people with different degrees of cognitive and physical disabilities will be able to have a more independent and easy life. Little children and babies will also be cared for in a more secure way while their parents are away. Thus, dependence on special caregivers will be decreased.

There are several prototypes, as well as commercially available products. When several applications are investigated, it is observed that these applications have common properties. Most of the existing solutions include one or more types of sensors carried by the patient, forming a body area network (BAN), and one or more types of sensors deployed in the environment, forming a personal area network (PAN). These two are connected to a backbone network via a gateway node. At the application level, the health-care professionals or other caregivers can monitor the vital health information of the patient in real-time via a graphical user interface (GUI). Emergency situations produce alerts and these alerts and other health status information can be reached via mobile devices like laptop computers, personal digital assistants (PDAs), and smart phones. A simple wireless sensor network application in health care is depicted in Figure 2.4. Based on this observation, in a typical scenario.

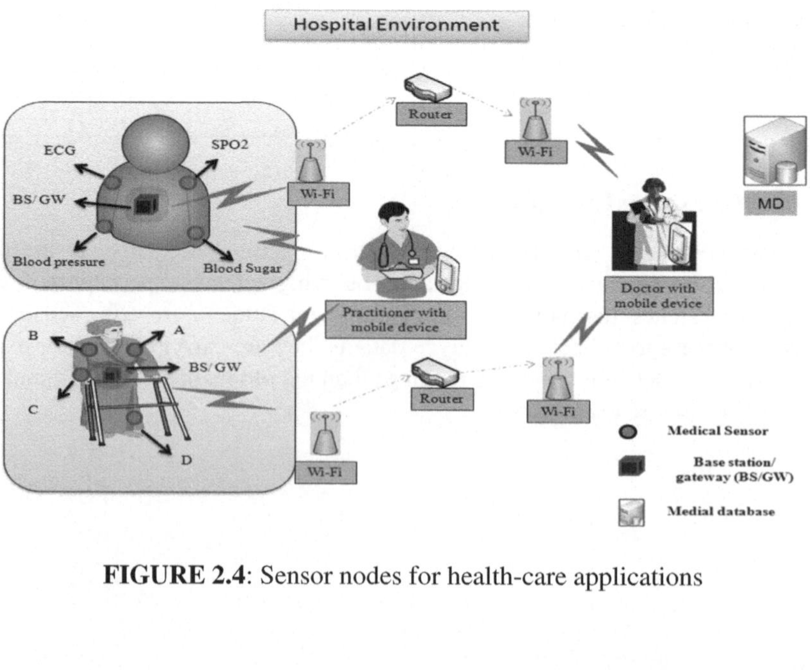

FIGURE 2.4: Sensor nodes for health-care applications

2.11 Underground coal mining

Frequent coal-mining accidents are a very significant problem for many mines. Thousands of miners die due to unawareness of the irregular methane concentration, temperature, or strain under the coal mines. These will be avoided if a monitor system can be developed to measure multi-parameters at many locations under the mine in real-time. Sensor nodes are deployed in mines to detect the presence of methane or other poisonous gases. After detecting the gas, the sensor nodes immediately transmit the message to the outside control room where the exact locations of the mines is determined. Now, the control room can inform to the miners to avoid these regions. Sometimes the sensor nodes are placed within the helmet of the miners, and a few gateway nodes are deployed over the mines. Thus, the position the miners can be tracked from the outside control room.

2.12 WSNs and IoT

Over the years, device-to-device communication or device-to-human communication has played an important role. These devices are used for personal electronics, home appliances, health monitoring, smart vehicles, and industrial applications. The use of these devices with the help of Internet connectivity has evolved the platform known as the Internet of Things (IOT). Almost all the devices which are used for health monitoring, industrial automation devices, vehicles and home appliances now possess Internet connectivity capability, which has increased the adoption of the IoT [45]. However, TCP/IP has some limitations in achieving device-to-device communication and low power wireless personal area networks (6LoWPAN) play the vital role in this regard. 6LoWPAN has become a standard for device-to-device communication technology in the IoT. Much research has been carried out on the use of WSNs in the IOT. Kruger et al. [46] have designed a gateway device that is integrated with wireless sensor networks. The device is used for monitoring the water consumption in a building. Therefore, the devices were integrated together in order to create an IoT featuring wireless sensors and laptops, smart phones and personal computers. The gateway communicated between 6LoWPAN-based intelligent water meters and IPv6 LAN networks in the building.

Chapter 3

A Survey

Chapter 3

A Survey

There is a rich literature available on the various ways for reducing energy consumption of wireless sensor networks. The research on these areas has been focused on system-level power awareness, such as dynamic voltage scaling, radio communication hardware, low duty cycle, and energy-aware MAC protocols [4, 9, 10, 11, 47, 48, 49, 50, 51]. At the network layer, the main aim is to find ways for energy-efficient clustering and routing of data from the sensor nodes to the base station so that the lifetime of the network is maximized [52, 53, 54, 55, 56, 57, 58, 59, 60]. As we are concerned with energy-efficient clustering and routing algorithms, we will perform a review of existing clustering and routing algorithms only by categorizing them into load balancing, metaheuristic, and distributed approaches.

3.1 Approximation algorithms

Low et al. [32] have proposed two different clustering algorithms. Their first algorithm, called the load balanced clustering algorithm (LBCA), assumes a special case in which the traffic load contributed by all sensor nodes is equal. To form a cluster, LBCA constructs an individual BFS (breadth first search) tree for each sensor node to find out the least loaded CH for assigning the sensor node to that CH. To assign a sensor node s_i, LBCA constructs a BFS tree rooted at s_i. Then the set of gateways onto which the sensor node s_i may be assigned is included into the tree as child of s_i. The BFS tree rooted at s_i is now constructed level-by-level. The tree is next extended to the third level by including the set of sensor nodes that were already assigned to the gateways in level 2 of the tree. It can be observed that the tree alternates between sensors and gateways from one level of the tree to the next. The sensor nodes are always in an odd level and gateways are in an even level. After constructing the BFS tree, the algorithm finds a path P from s_i to the gateway (say g_k) having the least load in the BFS tree. Let P be denoted by: $s_i \rightarrow g_2 \rightarrow s_3 \rightarrow \cdots \rightarrow s_{k-1} \rightarrow g_k$, where sensor s_3 was already assigned to gateway g_2 and s_i can be assigned to gateway g_2. Now, all of the sensors from the path P are next reassigned to the gateway as follows:

i) Assign sensor s_i to gateway g_2.

ii) Next, reassign sensor s_3 to gateway g_4.

iii) In general, reassign sensor s_{2i-1} to gateway g_{2i}, where $i = 2, 3, \cdots, \frac{k}{2}$.

It is shown and proven that the LBCA produces an optimal solution for the load-balancing problem with equal load of the sensor nodes. However, the time complexity of LBCA is $O(mn^2)$ for n sensor nodes and m gateways, which may not be effective for large-scale WSNs. Moreover, for a large WSN, building a BFS tree for individual sensor node takes a substantial amount of memory space.

Authors have also proposed an approximation algorithm called GLBCA (greedy load balanced-clustering algorithm) for the case where sensor nodes have unequal loads. To form a cluster, GLBCA considers a bipartite graph of the sensor nodes and the CHs to find out the maximum matching for assigning a sensor node to a CH. The algorithm is shown to be a 1.5-approximation and the time complexity is $O(n[n + m + q])$ where q is the cardinality of the bipartite graph.

3.2 Centralized algorithms

Although clustering of sensor nodes is proven to be more energy efficient than the traditional WSNs, the cluster-based architecture is very much dependent on the cluster heads. The CHs die quickly due to extra workload. Moreover, with imbalance of the load, some overloaded CHs may die faster than the other CHs. As a result, precious data of a particular region of the network is not able to reach to the base station; even the network may be partitioned or disconnected. Therefore, load balancing of the cluster heads is a crucial issue, that has been studied by many researchers. Here, we review some algorithms that are centralized in nature.

Gupta et al. [38] have proposed a load-balanced clustering called LBC, where the authors define cardinality of a cluster as the number of sensor nodes associated with the cluster and attempts to minimize the variance of the cardinality of each cluster in the system. LBC takes $O(mn \log n)$ time in the worst case for n number of sensor nodes and m number of CHs in the networks. The authors show, in [61], that suitable clustering techniques can be used to increase the system's lifetime.

In [39], the authors focus on fault-tolerant clustering, and propose a two- phase fault-tolerant approach, namely, detect phase and recover phase. The types of failures considered in [39] include complete failure, link failure, and range failure. The idea is to perform periodic checks on the status of the gateway nodes so that the system can learn about the failure of any gateway node. The clustering scheme includes the creation of backup information during the clustering phase, which can be used to re-assign sensor nodes managed by the any failed gateway node, thereby eliminating the necessity of a full-scale re-clustering involving the entire network.

In recent years, many clustering and routing algorithms have been developed following the centralized approach. The authors in [62] have reported an algorithm called CELBA (centralized energy-efficient load-balanced algorithm). The algorithms presented in [62] and [38] assume that all the sensor nodes generate equal traffic load to their CHs. Unfortunately, the algorithms may not be effective for the network that generates unequal traffic load to the CHs. CHLBC [63] is a load-balancing algorithm for wireless sensor networks. After the selection of cluster heads, it constructs the backbone network, which is composed of cluster heads and builds up the multi-hop transit routes from cluster heads to the BS. CHLBC adjusts the size of the cluster according to the relay traffic of the cluster head. The ordinary nodes select the cluster head that is closer and has small relay traffic. CHLBC creates the overhead of control packets during the process of construction of hop field and calculation of relay traffic. Load balancing group clustering (LBGC) [64] is a clustering and routing algorithm which periodically selects cluster heads and implements dynamic route calculation according to the condition of energy distribution of the network, which could make full use of the heterogeneous energy to realize load balance. It presents a multi-hop group formation method with unequal size to balance the energy consumption among the groups. LBGC reselects the cluster head in each group considering its communication costs and surplus energy, and reorganizes the cluster with a new cluster head from the group. However, it is applicable in the scenario where sensor nodes are uniformly distributed. Li et al. [65] have presented an algorithm to find the best location of the cluster heads for minimizing energy consumption of the networks. However, re-clustering or routing complicates the network setup in the centralized approaches. A new communication schedule and network setup are required to be communicated to the whole network, which may not be realistic for a large area WSN.

3.3 Metaheuristic algorithms

Finding optimal load-balanced clusters and route is shown to be NP-hard problem. With the proliferation of soft computing techniques, metaheuristic algorithms have been studied to solve many optimization problems when the computational complexity of finding the optimal solution through a brute force approach is very high. A number of metaheuristic based clustering and routing algorithms have been reported for WSNs [66].

3.3.1 Genetic algorithm-based routing by Bari et al.

Bari et al. [31] have proposed a GA-based algorithm for data routing between gateways in a two-tier wireless sensor network. Chromosomes are represented as a string of relay node numbers. Length of the chromosomes is kept the same as the relay node number. The fitness value is computed as the lifetime of the network, which is represented by the total number of rounds until the first relay node runs out of battery power. The fitness function for an individual is computed as $L_{Net} = \frac{E_{Initial}}{E_{Max}}$, where L_{Net} denotes network life in terms of rounds and $E_{Initial}$ is the initial energy of a relay node. It is assumed that the value of $E_{Initial}$ is known beforehand, and that it is the same for all relay nodes. E_{Max} is the maximum energy dissipated by any relay node in one round of data gathering.

Here, selection of individuals is carried out using the Roulette-wheel selection method. For mutation operation, a critical node is selected from the relay nodes, which dissipates the maximum energy due to receiving and/or transmitting data. Mutation is done by either replacing the next-hop node of this critical node by a new next-hop relay node or by diverting some incoming flow towards that critical node to another relay node.

3.3.2 GAR

In GA-based routing (GAR) [30], the overall communication distance from the gateways to the BS is minimized. A routing schedule for a sub-graph network along with the corresponding chromosome of eight relay nodes and one BS is shown in Figure 3.1.

This is a directed acyclic graph $G(V, E)$, where E represents the edges, C_i is a relay node, and C_j is either a relay node or a BS such that C_j is closer to the BS than C_i and C_j is within communication range of C_i. Figure 3.1(b) is the corresponding chromosome. In this example, the value of the gene in position 1 is 6, indicating that node 1 selects the node 6 for data transmission. The value at 6 is 5, indicating that 5 is the next selected relay node, and the value at 5 is 9, indicating the next relay node, which is the BS (destination). Therefore, the route is expressed as the path $1 \longrightarrow 6 \longrightarrow 5 \longrightarrow 9$.

However, it is different from [31] with respect to the following issues. For selection of individuals, tournament selection is used in contrast to Roulette-wheel selection. Fitness function is defined in terms of total distance covered in a round rather

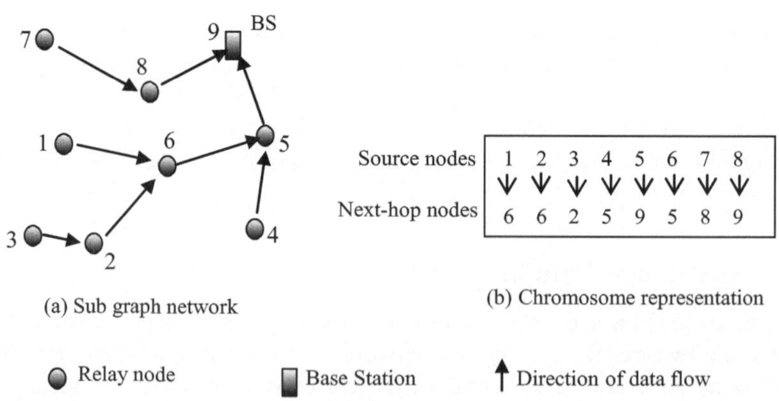

FIGURE 3.1: Sub-graph network of WSN and the corresponding chromosome representation

than network lifetime in terms of number of rounds. In the mutation phase, we select a relay node that uses maximum distance to transmit the data to its neighbor in contrast to a critical node as defined in [31]. However, both of the algorithms presented in [31] and [30] consider only routing of aggregated data from the gateways to the BS without considering data communication from the sensor nodes to the gateways within each cluster.

3.3.3 GA based hierarchical clustering by Sajid et al.

Sajid et al. [67] have presented a GA based hierarchical clustering algorithm to choose a set of cluster-heads from the normal sensor nodes. For the fitness parameter they have used i) direct distance, ii) cluster distance, iii) cluster distance-standard deviation, iv) transfer energy and v) number of transmission. However, their method selects CHs from normal sensor nodes and form clusters without any load balancing. As a result, some CHs may quickly die and requires frequent re-clustering.

3.3.4 Energy aware evolutionary routing protocol (EAERP)

Enan et al. [68] have presented an energy-aware evolutionary routing protocol (EAERP) for dynamic clustering of wireless sensor networks. Here the authors have made an attempt to minimize the energy consumption throughout the network by choosing a set of efficient cluster heads from the normal sensor nodes while all non-CH sensor nodes determine the nearest CH to join. A complete clustered route solution is presented as an individual (or chromosome). The length of each chromosome is taken as being the same as the number of sensor nodes in the network (say N). The gene values of a chromosome can be either 0 for non-CH nodes, 1 for CH nodes, or -1 for dead nodes (i.e., inactive sensor nodes having no energy). A population of n chromosomes, $I_n = \{I^i_j | \forall i \in \{1, 2, \ldots, n\} \text{ and } \forall j \in \{1, 2, \ldots, N\}\}$, can be formally specified as

$$I^i_j = \begin{cases} 1 & \text{if } E_R(s_j) \geq 0 \text{ and } s_j = CH, \\ 0 & \text{if } E_R(s_j) \geq 0 \text{ and } s_j = non - CH, \\ -1 & \text{otherwise} \end{cases} \tag{3.1}$$

Each individual is randomly initialized with 1 and 0 according to the probability p as used in LEACH by the equation (3.4.5).

But, EAERP suffers from the same problem as LEACH [19] faces, as some sensor node may become a CH that may not have sufficient energy. Moreover, EAERP requires re-clustering in each round to rotate the extra workload of CHs. Unfortunately, being a centralized approach, EAERP requires the entire network information in each round for re-clustering.

Jin et al. [69] have presented a method to find the CHs based on a fitness function that minimizes the total transmission distance in the network. Here, GA is used to select only cluster heads. Each non-CH sensor node uses a deterministic method to find its nearest cluster-head. All the methods presented in [70] and [71] also consider only the Euclidian distance between the cluster heads or BS, and they did not consider any other parameters like residual energy, load balancing, or standard deviation. Mudundi et al. [70] have proposed a genetic clustering algorithm (GCA) for dynamic formation of clusters in WSNs with the goal of increasing the lifetime of the network by minimizing the energy dissipation. The fitness function is developed by using the number of CH nodes and the Euclidian distance between all the nodes in each cluster to their CH with some weight value. Yang [71] has presented a method called harmony search algorithm (HAS), for increasing the lifetime and reducing the energy consumption in a cluster-based WSN. In the fitness function, they have used the maximum Euclidean distance of the nodes, and the ratio of the energy of all the sensor nodes with the total current energy to the CH in the current round. In [72], the authors have presented a clustering method to optimize the energy consumption of the sensor nodes. Their method of CH selection is based on several parameters, such as residual energy, communication energy, number of CHs and the distance between CHs, and their member sensor nodes. However, all the above algorithms [69, 70, 71, 72] do not consider any load balancing of CHs through proper formation of clusters.

BCEE [73] aims at equalizing the consumption of energy in sensor nodes, and to prolonging the network lifetime, by adopting a mixed hierarchical and ant-based routing method. The protocol operates in two phases similar to other hierarchical protocols where routing is done in two phases. In phase one, cluster heads are selected adopting a K-means algorithm where the actual location of the nodes is unknown, but it uses the idea of receive signal strength indicator (RSSI) for cluster formation. And in the second phase, called the steady state, this method utilizes the techniques of ant colony optimization to establish an optimal multi-hop route from cluster heads to the sink using a rational and hop-selecting technique. The protocol was compared with LEACH in simulations and performs better in terms of energy consumption of sensor nodes. BCEE has some drawbacks in terms of network overhead, as cluster formation also requires more energy in the network, and delay of data transmission to the sink node.

Wang et al. [74] have proposed an energy-balanced ant based routing protocol (EBAB) that is an adaptive dynamic routing algorithm based on ant colony optimization. The algorithm is divided into two parts, intra-cluster and inter-cluster. In the intra-cluster, the algorithm is further divided into rounds, and clusters created at the beginning of the routing process. The intra-cluster comprises the selection of cluster heads where cluster heads are in competition based on the areas, the sensor node belongs to. After the cluster heads selection, cluster is set up, and cluster head sends a message in a range to tell the other sensor nodes. The sensor nodes send an ACK message to join the cluster head. If the node receives more than one message, it will choose the optimal cluster head according to the distance and energy. When the cluster head energy falls below a certain value, it broadcasts information about its remaining energy, and the distance of other cluster members. Then after calculating the fitness of all the other sensor nodes, the node that has the biggest fitness will become the successor of the cluster head. The intra-cluster section uses the improved ACO algorithm. When compared with LEACH, it performs better in terms of survived number of nodes and the amount of packets received at the sink, at the end of the simulation.

Chakraborty et al. [75] have presented a differential evolution-based routing algorithm for more than a thousand relay nodes, such that the energy consumption of the maximum energy-consuming relay node is minimized. However, the authors do not take care about the cluster formation. Some improper clustering may lead to serious energy inefficiency of the relay nodes.

Latiff et al. [76] have proposed a centralized particle swarm optimization based clustering (PSO-C). The authors consider both energy available to nodes and physical distances between the nodes and their cluster heads. Each particle represents a combination of cluster heads. The fitness function for the PSO-C is defined as $f = \beta \times f_1 + (1 - \beta) \times f_2$, where f_1 is the maximum average Euclidean distance of the nodes to their associated cluster heads and f_2 is the ratio of total initial energy of all nodes to the total energy of the cluster head candidates. This ensures that only the nodes that have above average energy resources are selected as the cluster heads, and that the average distance between the nodes and the cluster heads is minimum. However, the algorithm considers only CH selection. It does not take care of proper cluster formation and load balancing of the CHs. Singh and Lobiyal [77] have also used the PSO for CH selection among the normal sensor nodes, and do not take care of the cluster formation. PSO and ant colony optimization (ACO) are used in WSNs for other optimization problems as well and they can be found in [58, 78, 79, 80, 81, 82, 83, 84, 85].

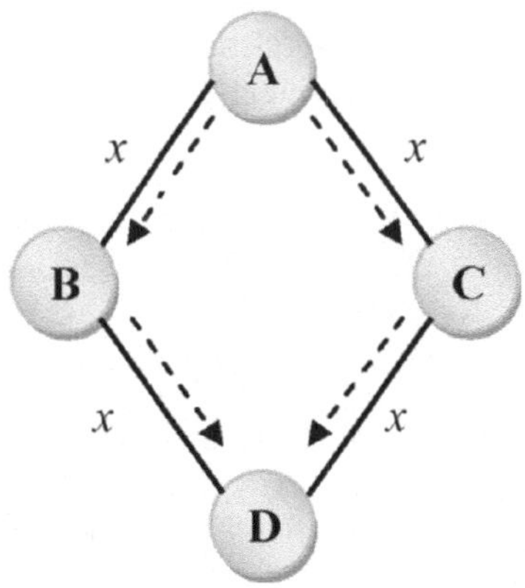

FIGURE 3.2: The implosion problem. Node *A* starts by flooding its data to all of its neighbors, *B* and *C*. *D* gets eventually two identical copies of data, which is not necessary the same

3.4 Distributed algorithms

All the above methods follow the centralized approach, which is useful for WSNs where location of each sensor node is known to the central controller or base station. Moreover, being a centralized approach, they are time consuming and catastrophic in event that the base station fails. Therefore, a number of distributed algorithms have been reported for clustering and routing. We describe a few popular distributed clustering and routing algorithms as follows.

3.4.1 Flooding

Flooding is an old technique to route data in ad-hoc wireless networks [86, 87]. Flooding is a reactive routing technique, and it does not require any routing algorithm, costly topology maintenance of the network, and complex route discovery algorithms. It can also be used for routing in sensor networks. In flooding, after receiving a data packet, each sensor node broadcasts the packet to all of its neighbors and this process continues until the packet reaches the destination node or travels for maximum number of hops. Although flooding is very easy to implement, it has many deficiencies, such as:

- *Implosion*: This is a situation where duplicated messages are sent to the same node. For example, as shown in Figure 3.2, sensor node A has two neighbor sensor nodes, B and C which are also the neighbors of sensor node D. The sensor node D receives two identical copies of the message sent by sensor node A.

- *Overlap*: If two nodes share the same observing region, both of them may sense the same information at the same time. As a result, neighbor nodes receive duplicated messages. As shown in Figure 3.3, B is getting redundant data y through C and D which are covering an overlapping geographic region.

- *Resource blindness*: The flooding protocol does not take into account of the available energy resources. An energy resource aware protocol must take into account the amount of energy available to the nodes at all time.

3.4.2 Gossiping

Gossiping is a derived version of flooding [87]. Here, after receiving the data packet, sensor nodes do not broadcast the packet to their neighbors, but send the packets to a randomly selected neighbor. Once the neighbor node receives the data, it randomly selects another sensor node to forward the packet to, and so on. Gossiping avoids the problem of implosion by just selecting a random node to send the packet rather than broadcasting. However, this causes delay in propagation of data through the nodes.

3.4.3 Sensor Protocol for Information via Negotiation(SPIN)

SPIN is the first data-centric adaptive routing protocol where, the sensor nodes broadcast an advertisement for the available data and wait for a request from interested sinks [86]. It is designed to address the deficiencies of classic flooding by negotiation and resource adaptation. In SPIN, data are named using meta-data. The meta-data negotiation helps in elimination of overlapping, redundant information, and resource blindness as drawbacks of flooding and thus saves a lot of energy.

This protocol uses three types of messages: ADV, REQ, and DATA. When a sensor node detects an event, it advertises the event to its neighbor sensor nodes by an ADV message containing a descriptor, i.e., meta-data, of the event as shown in Figure 3.4a. The neighbor sensor node that is interested in the data sends an REQ message and DATA message is sent to the neighbor sensor node as shown in Figure 3.4b and Figure 3.4c, respectively. The neighbor sensor node then repeats this process, as illustrated in Figure 3.4d-Figure 3.4f. As a result, the sensor nodes in the entire sensor network, that are interested in the data, will get the copy.

SPIN does not ensure the delivery of relevant data to all interested nodes or even to the sink node. For instance, if the nodes that are interested in the data are far away from the source node, and the nodes between source and destination are not interested in that data, such data will not be delivered to the destination at all. Therefore, SPIN is not a good choice for applications such as intrusion detection, which require reliable delivery of data packets over regular intervals [52].

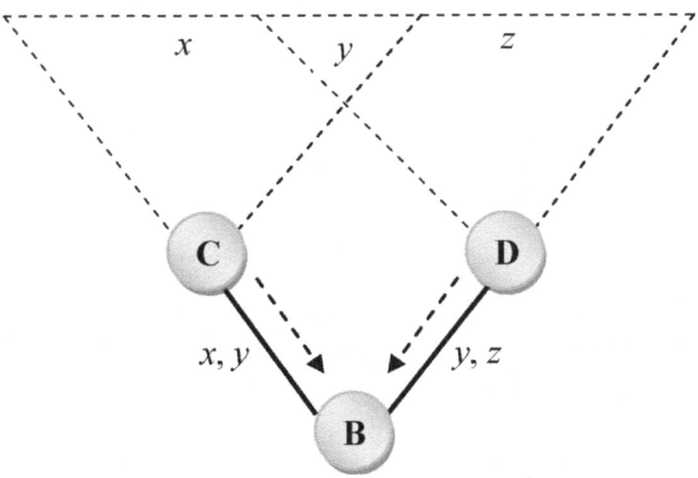

FIGURE 3.3: The overlap problem. Two sensor nodes *C* and *D* cover an overlapping geographic region and *B* gets the same copy of the data from these sensors.

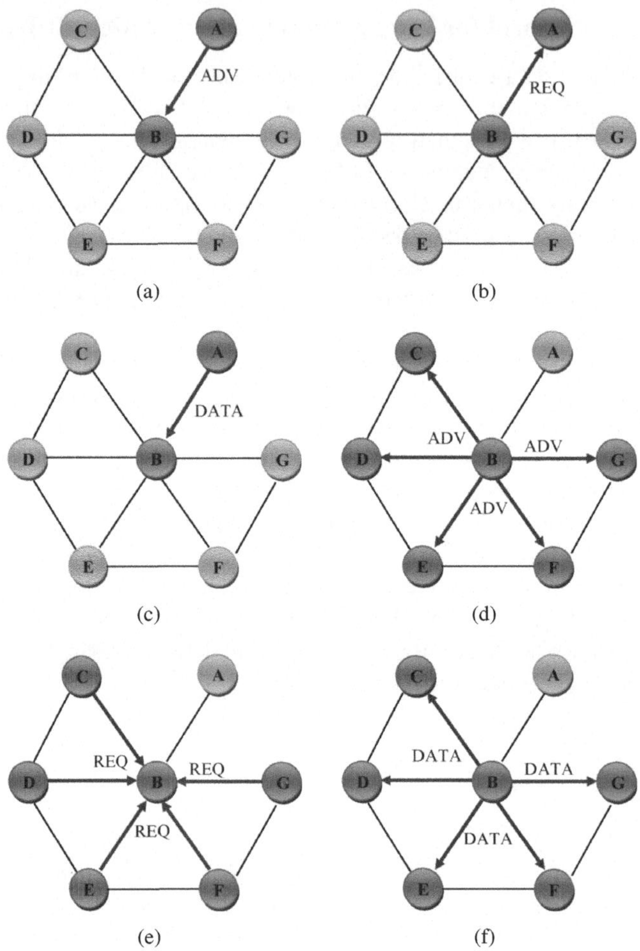

FIGURE 3.4: SPIN protocol. Node *A* starts by advertising its data to node *B* (a). Node *B* responds by sending a request to node *A* (b). After receiving the requested data (c), node B then sends out advertisements to its neighbors (d), who in turn send requests back to *B* (e-f).

3.4.4 Directed Diffusion (DD)

Directed diffusion (DD) is a data-centric routing scheme where the sink broadcasts the interest queries [88, 89]. The idea is to diffuse the data through sensor nodes by using a naming scheme for the data. Direct diffusion uses the attribute-value pairs for the data and queries the sensors on an on-demand basis by using the attribute-value pairs. Here, the sink node sends an interest query to all sensor nodes. The query is created using a list of attribute value pairs, such as name of event, inter-

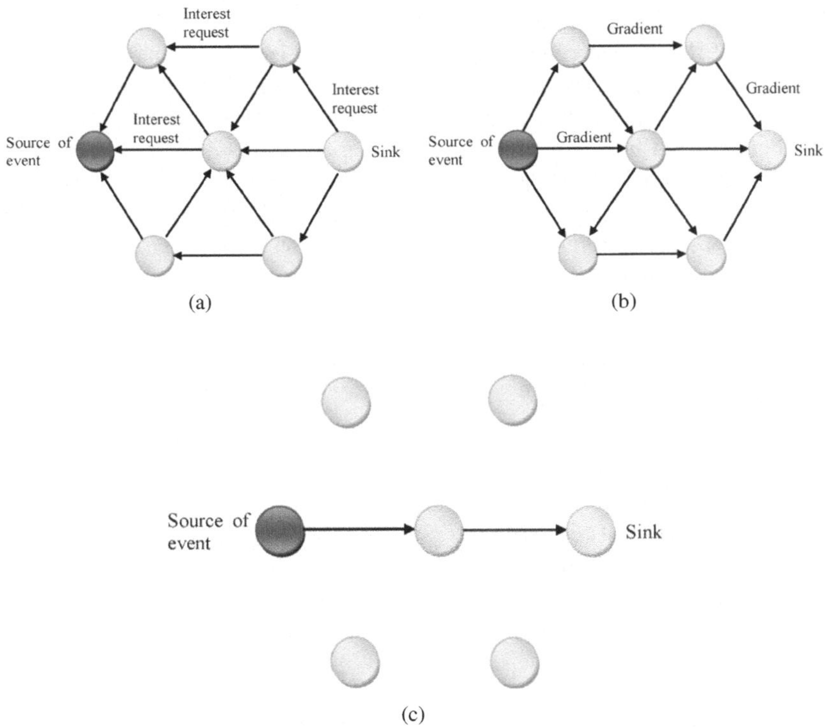

FIGURE 3.5: Directed diffusion. (a) Interest propagation, (b) initial gradients setup, and (c) data delivery along reinforcement.

val, duration of the event, geographical location, etc. Each sensor node then stores the query entry in its memory and is propagated throughout the sensor network. The stored query entries are then used to compare with the received data with the values. When some source node has data for the given attribute-value, the source node sends the data to the sink node.

The interest query contains several gradient fields. The gradient direction is set to the neighboring node from which the interest is received. Hence, by utilizing the gradients, paths are established between source and sink as shown in Figure 3.5. Obviously, several paths are established between sink and source and the sensing data starts flowing toward the sink by using multiple gradient paths. One of the established paths is selected by reinforcement. The sink resends the original interest message through the selected path with a smaller interval, hence reinforcing the source node on that path to send data more frequently. Path repairs are also possible in directed diffusion. When a path between a source and the sink fails, a new or alternative path is identified. For this, directed diffusion basically re-initiates reinforcement by searching among other paths [2, 52].

Directed diffusion differs from SPIN in terms of its on-demand data querying mechanism. Here, the sink queries the sensor nodes by flooding if a specific data is available. In SPIN, the sensor nodes themselves advertise the availability of data and allow the interested nodes to request for the data. Since, directed diffusion is data centric and on demand, it has many advantages. All communication is neighbor-to-neighbor with no need for a node-addressing mechanism. It is energy-efficient and there is no need for maintaining global network topology. However, directed diffusion cannot be applied to all sensor network applications as it is based on a query-driven data delivery model. Applications like environmental monitoring which requires continuous data delivery to the sink, will not work efficiently with this protocol [2, 52].

3.4.5 Low-energy adaptive clustering hierarchy (LEACH)

Low-energy adaptive clustering hierarchy (LEACH) [18, 19] is a popular distributed cluster-based routing algorithm. The operation of LEACH is separated into two phases, the setup phase and the steady state phase. In the setup phase, the clusters are organized and cluster heads are randomly selected. In this phase, each sensor node s_n generates a random number between 0 and 1 and compares it to a pre-defined threshold $T(n)$. If a random number be less than $T(n)$, then the sensor node decides to be CH in that round, otherwise it will be a cluster member. The threshold $T(n)$ is defined as follow:

$$T(n) = \begin{cases} \dfrac{p}{1 - p \times (r \bmod (\frac{1}{p}))} & \text{if } s_n \in S, \\ 0 & \text{otherwise.} \end{cases} \tag{3.2}$$

where p is the desired percentage of CHs in the network, r is the current round, and S denotes the set of sensor nodes that have not been selected as a cluster head in the last $\frac{1}{p}$ rounds. Once sensor nodes have elected themselves to be CHs, that broadcast an advertisement message using a non-persistent-CSMA MAC protocol to inform their neighbor about the selection. Each non-CH sensor node selects its CH for this round only based on the received signal strength of the advertisement, i.e., a sensor node joins the CH that is closer to it. The CHs set up a TDMA schedule and transmit this schedule to their cluster members. The TDMA schedule ensures that there are no collisions among data messages inside a cluster.

In the steady state phase, the actual data is directly transmitted to the BS. Data is sent from the CHs to the BS using a fixed spreading code and CSMA. It dynamically and randomly rotates the workload of the CHs among the sensor nodes, which is useful for load balancing. The positive aspect of random selection of cluster heads is that the sensor nodes will unevenly deplete their energy, and therefore they should randomly die throughout the network. Since the clustering implemented in LEACH is based on randomness; its cost is much less and realistically feasible.

Although the LEACH clustering protocol seems to be promising, further evaluation is needed. The method has certain disadvantages. Firstly, there is a chance of selecting a node with low energy as a CH, as the selection is made randomly. Secondly, the CHs use single-hop communication for data forwarding directly to the base station and therefore consume more energy. Also it is impractical for WSNs with a large coverage area. A large number of algorithms have been developed to improve LEACH and can be found in [90, 91, 92, 93, 94, 95].

3.4.6 Efficient Routing-LEACH (ER-LEACH)

Al-Refai et al. [96] have proposed efficient routing-LEACH (ER-LEACH). In this protocol, the authors have focused on the data that is requested by the BS from the environment. Therefore, data can be aggregated by sensor nodes. In this proposal, the authors have removed some of the problems in the classical LEACH based protocol. The ER-LEACH protocol enhances the lifetime of WSNs through load balancing using the following steps:

- By enhancing the selection of the CH during the setup phase.
- By using an alternative CH to take the role of the CH in case the underlying CH dies.
- By using zone routing protocol for load balancing of WSNs.
- By performing well, especially when the mobility is very high.

3.4.7 Adaptive and energy efficient clustering (AEEC)

Buyanjargal and Kwon [97] have presented AEEC (Adaptive and energy efficient clustering), a modified algorithm of LEACH. Although AEEC performs a clustering phase similar to LEACH, it selects the CHs based on the remaining energy of the sensor nodes and thus overcomes the problem of random selection of CHs. However, the main drawback of this protocol is a high volume of control massage exchange between the selected CH and the sensor nodes to form a cluster.

3.4.8 Power-efficient gathering in sensor information systems (PEGASIS)

Power-efficient gathering in sensor information systems (PEGASIS) [98, 99], and its variation hierarchical-PEGASIS, are two improvements of LEACH. Rather than forming multiple clusters, PEGASIS forms chains of sensor nodes so that each sensor transmits and receives from a neighbor, and only one node is selected from that chain as group head to convey data to the BS. It assumes all the sensor nodes have a global knowledge of the network topology, which allows the use of a greedy algorithm when constructing a chain. The chain starts with the farthest node from the base station. In this approach, PEGASIS ensures that the nodes farthest away from the base station have an accessible neighbor. With each dying node, the chain is reconstructed; the group head of each round of communication will be at a random place selected in the chain. As the data moves from node to node in the chain, it is fused together. Eventually the designated node in the chain will transmit the fused packet to the base station. However, it requires dynamic topology adjustment and the data delay is significantly high, which is unsuitable for large scale networks.

3.4.9 Hybrid energy efficient distributed clustering (HEED)

Hybrid energy efficient distributed clustering (HEED) [100] is a distributed clustering protocol that uses a hybrid combination of the residual energy and the intra-cluster communication cost as attribute for cluster head selection. HEED ensures a uniform distribution of cluster heads across the network and adjusts the probability of cluster heads selection to ensure inter cluster heads connectivity. To prevent sensors from having more than one cluster-head, this algorithm introduces an inter-cluster communication energy cost, and sensors choose a cluster-head to minimize the energy cost. However, to balance residual energies of sensors, HEED requires frequent re-clustering causing a high, overhead cost, and synchronization problem among sensors. Moreover, it requires larger time delay due to iterative communications among sensor nodes for cluster head selection.

3.4.10 TEEN and APTEEN

TEEN [101] is a hierarchical protocol whose main aim is to respond to sudden changes in the sensed attributes, such as temperature. This protocol combines the hierarchical technique in line with a data-centric approach. It then involves the formation of clusters along with cluster leaders that broadcast two thresholds to the nodes; the hard and soft thresholds. Hard threshold have the minimum values of an attribute for a sensor node to trigger to power on its transmitter so as to transmit to the cluster head. It is normally not suited in applications where continuous data is needed, since it is threshold dependent.

APTEEN [102] is an improved version of TEEN, whose main function is not limited to the formation of clusters, but also aims at both capturing periodic data and reacting to time dependent events. In APTEEN, cluster leaders perform aggregation as well as conserves energy. Three queries are supported in the protocol; historical for analysis of past information values, persistent for monitoring of events for some time duration, and one-time for snapshot views of the sensor network. Simulation results show that it outperforms LEACH. This is because of the overhead and complexity in clusters formation in multiple levels, and implementation of the threshold based functions in LEACH.

3.4.11 DEEC: Distributed energy efficient clustering

Li et al. [103] have proposed a distributed energy-efficient clustering algorithm called DEEC for heterogeneous networks. In DEEC, the cluster-heads are selected by a probability based on the ratio between the residual energy of each node and the average energy of the network. However, the exact information of the total energy of the network is difficult to obtain due to a high node failure rate or redeployment. Therefore, DEEC cannot be applied in a harsh or hostile environment.

3.4.12 DEBR: Distributed energy balance routing

DEBR [104] is a distributed energy balance routing scheme in which a new heuristic metric is suggested. The algorithm uses a path with energy sufficiency as well as energy efficiency to pursue energy balancing of the sensor network. Energy sufficiency depends upon the available energy, and energy efficiency depends upon the required energy. By using a composite of both quantities, a good path that achieves energy balance can be found. DEBR defines a composite measure of the energy cost EC_{ij} to transmt from sensor node s_i to sensor node s_j as

$$EC_{ij} = \frac{\text{Required energy from node } s_i \text{ to } s_j}{\text{Available energy at node } s_i} = \frac{e_{ij}}{E_i^r} \tag{3.3}$$

The total energy cost (TEC_{ik}) of a neighboring node s_k at sensor s_i is the sum of the energy costs from node s_i to s_k and from node s_k to the base station (BS)

$$TEC_{ik} = EC_{ik} + EC_{k,BS} \tag{3.4}$$

Based on this definition, sensor s_i can select the best candidate relay node (say s_r) having minimum TEC value, i.e.,

$$TEC_{ir} = Min\{TEC_{ij}, \forall j \in Neighbor(s_i)\} \tag{3.5}$$

It is observed that DEBR occasionally prevents sensors from using the most energy efficient path to achieve an energy balance of the network. Moreover, it guarantees elimination of loops in any routing path and ensures finding a routing path to the base station without loops. Since the cost function EC_{ij} is a ratio of transmission energy cost and available energy, its value is low in the case that the required transmission energy is low and available energy is high.

The main drawbacks of DEBR are as follows: (1) It assumes each sensor node uses a fixed transmission power for communicating with its neighbor sensor nodes. The neighboring distance is the maximal reachable distance with the fixed transmission power for neighbors. Therefore, it is not considering the exact distances to neighbor nodes. (2) All sensor nodes make a local routing decision. Node s_i selects s_r as the best candidate relay node for transmitting data to the base station as per the equation (3.4) without considering whether s_r can send data directly to the base station or not. (3) In order to balance energy consumption of the network, a sensor node may select another sensor node as a next-hop relay node, which may not have any other next-hop within its communication range. Thus, it stops the dissemination of some precious data to the base station.

3.4.13 Exponential and Sine Cost Function based Routing

ESCFR (Exponential and Sine Cost Function based Routing) [105] is an energy aware routing algorithm. ESCFR improves the cost function of DEBR by considering the following principle: “A cost function should have the following properties: small changes in nodal remaining energy can lead to large changes in the function value."Keeping this principle in mind, authors have used exponential and sine functions which are the kind of functions where small changes in variables can cause large changes in function values. These two types of functions are combined together and construct an exponential function, as represented by equation (3.6).

$$f(x) = \exp(1/\sin(x)) \tag{3.6}$$

Now the exponential function is used to derive an efficient cost function where the current remaining energy E_i^r of a sensor node s_i is used as follow:

$$EC_{ij} = e_{ij} \times \exp(1/\sin(\pi - \frac{\pi}{2}\frac{E_i^r}{E_0})) \tag{3.7}$$

where E_0 is the initial energy of a sensor node. It is shown that the energy-balancing performance of exponential and sine cost function based routing (ESCFR) is no worse than DEBR. However, ESCFR suffers from the same problem as DEBR. Other clustering and routing algorithms developed for WSNs can be seen in [106, 107, 108, 109, 110, 111, 112, 113, 114, 115, 116, 117, 118].

3.5 Mobile sink routing (MSR)

We now show a review of the available routing procedures for a mobile sink. Scanty and detached WSNs can be preferably supervised by an MS [52] where the MS might acquire the sensory intelligence from unreachable parts of the system that may be not available in a scenario with a stationary base station, hence improving the availability of the system. MSR conventions can collect data from freely associated regions of the system that disburse unnecessary assets to adapt to such scenarios [119]. In delay-tolerant applications, a moving sink likewise lessens the number of hops on data transmission paths or courses. In such applications, information accumulation points are used that hold up and disperse information when the sink visits nearer at a price of extended data latency (please refer to Figure 9.2). The shorter information transmission routes lead to an enhanced amount of data passing through a system, and dependability simultaneously with diminished power expenditure. The SNs to the BS transmission might be taken care of by a grouping or clustering strategy. As a matter of first importance, the data propagation (or routing) issue requires a collaboration plan since the system is not being centrally controlled, and the passage accessibility of SNs alter progressively. The sensory intelligence must be propagated from the phenomenon area to the BS in such situation. The SNs are grouped to form a cluster and collaborate among one another for the MSR and the passage allocation. Information passage can be controlled by the group leader and group associates for intra-group associations [120].

There are several ways to deal with the issue of routing in WSNs with MS; the hierarchical mobile sink routing (HMSR) procedures are the most vital and commonly opted for. These solutions are mean to diminish the burden of broadcasting the sink′s location to the system by setting up a virtual order of SNs that forces diverse lively acts on the SNs. The obtained hierarchy may be produced on two or more levels. The SNs in the arrangement (high-level SNs) get the sink′s location and the rest of the SNs (first-level SNs) query these high-level SNs to obtain the sink′s location statistics at any time. A fruitful hierarchical methodology ought to utilize an effectively reachable virtual high-level formation and advocate remedies to counter viable hop spots on the high-level SNs. In this segment, the most important MSR procedures are inspected and categorized as follows.

- **Grid-based procedures:** Conventions of this class utilize a grid formation. Different structures could be utilized to form the grid: hexagons, rectangles, triangles, and so on. As the grid is normally a geometric formation, physical coordinates of the SNs are needed, thus location-vigilant SNs are favored. The available literature can be found in [121].

- **Clustering based procedures:** Procedures having a place with this class utilize grouping methods to segregate the system and utilize the cluster head sensor nodes [122].

- **Overlay construction based procedures:** The procedures of this category build up an overlay foundation including all the SNs that are normally comprised of SNs along diverse acts [123].

- **Tree formation based procedures:** This category of procedures build a veneering virtual tree formation. The sink′s location is promoted from top to bottom (root-to-leaves) [124].

- **Geometric characteristics based procedures:** As SNs are normally placed over a target area that can be assumed in a 2D plane, geometric characteristics may be used to guarantee the meeting point of various sorts of intelligence. The justification behind this class of procedures is to utilize these characteristics with a specific end goal of acknowledge sink′s location to the SNs [125].

- **Area partition based procedures:** These methodologies assign the SNs in a territory of particular limits as the high-level SNs (nodes that are designated storage nodes) as opposed to setting up entangled formations. To alleviate the hot spot anomaly, or maybe then change the formation, the span of the territory is indicated as being sufficiently substantial to spread and diminish the additional load on the high-level SNs [126].

- **Rendezvous points based procedures:** The solutions in this category employ a few SNs (selected) as rendezvous points (RPs). These RPs must be in the communication range of the mobile sink trajectory (MST) or mobile sink (MS) and are utilized for intermediary buffering of sensory intelligence until MS visits them to collect the stored (or buffered) data [127].

Chapter 4

Approximation Algorithms for Clustering

Chapter 4

Approximation Algorithms for Clustering

4.1 Introduction

There are many optimization problems that are NP-complete or NP-hard in nature and as such there is a polynomial time solution for them. However, for such problems, we do not require an exact/optimal solution for many practical applications. Approximation algorithms are some alternative approaches that provide approximate or near optimal solutions in polynomial time. Approximation algorithms are used for many optimization problems where exact polynomial-time algorithms are known but are too expensive due to the input size [128]. Some popular examples of such problems for which approximation algorithms have been developed include vertex cover, set cover, jon scheduling, bin-packing, and travelling salesperson problems. Ideally, the solution may go optimal up to a small constant factor, say within 5% of the optimal solution. The ratio between the result obtained by an approximation algorithm and the optimal result is called the approximation ratio of the algorithm. An algorithm with approximation ratio k is called a k-approximation algorithm. For example, an approximation algorithm that results a solution of cost 2 in solving a given problem that has an optimal cost 1 has approximation ratio 2. Therefore, the algorithm will be called a 2-approximation algorithm. Clearly, the approximation ratio of an algorithm is greater than or equal to 1 and it is exactly equal to 1 if and only if it has got an optimal solution. Approximation algorithms attract huge attention for being used in many applications of wireless sensor networks, computational biology, transportation planning, and inventory management.

In this chapter, we discuss various approximation algorithms for load balanced clustering problem in wireless sensor networks which is defined as follows. Given a set of n sensor nodes and a set of m gateways, the problem is to assign all the sensor nodes to the gateways such that a sensor node can belong to one and only one gateway and maximum load of each gateway is minimized. By the load of the CHs, we mean the load contributed by the member sensor nodes due to data generation and communication. When the rate of data generation and/or rate of data transmission is the same for all the sensor nodes, then we say that sensor nodes contribute equal load to their CHs. On the other hand, when any or both of them varies, then the load will be considered unequal. It is also noteworthy that for a WSN with n sensor nodes and m gateways, the number of possible clusters is m^n. This implies that the computational complexity of finding the optimal load-balanced clustering for a large WSN seems to be very high by a brute force approach. In fact, load-balanced clustering with unequal load of the sensor nodes is a NP-hard problem [32, 129, 130, 131, 132, 133].

Here, first we present a load balanced clustering algorithm that produces an optimal solution in $O(n \log n)$ time in contrast to $O(mn^2)$ time, as required by the algorithm proposed by Low et al. [32] for a special case in which all sensor nodes have equal load. It is then shown that the same algorithm can work for the general case, i.e., sensor nodes having unequal loads. It is proven that this is actually a 2-approximation algorithm. Second, an improved algorithm is presented for the general case, which is 1.5-approximation. This algorithm is also shown to run in $O(n \log n)$ time. Finally, a load balanced clustering algorithm called HLBCA (heap-based load balanced clustering algorithm) is presented with $O(mn)$ time complexity. The algorithm uses max heap of sensor nodes to efficiently balance the load of the CHs. It is then shown that the algorithm produces an optimal solution for a special case in which all sensor nodes have equal load and is a 1.5-approximation algorithm for the general case, i.e., sensor nodes having variable loads.

4.2 System model and problem formulation

Here, a WSN model is assumed where all the sensor nodes are deployed randomly along with a few gateways, and once they are deployed, they become stationary. A sensor node can be assigned to any gateway if it is within the communication range of the sensor node. Therefore, there are some pre-specified gateways onto which a particular sensor node can be assigned. Thus each sensor node has a list of gateways and it can be assigned to only one gateway among them. Similar to LEACH [19], the data gathering operation is divided into rounds. In each round, all sensor nodes collect the local data and send it to their corresponding CH (i.e., gateway). On receiving the data, the gateways aggregate them to discard the redundant and uncorrelated data and send the aggregated data to the base station via another CH as a next hop relay node. Between two adjacent rounds, all nodes turn off their radios to save energy. All communication is over wireless link. A wireless link is established between two nodes only if they are within the communication range of each other. Current implementation supports TDMA [134] protocol to provide MAC layer communication. Gateways use slotted CSMA/CA MAC protocol to communicate with the base station [135]. The following notations are used for the problem formulation [136, 137, 138, 139].

- The set of sensor nodes is denoted by $S = \{s_1, s_2, \cdots, s_n\}$.
- The set of gateways is denoted by $\xi = \{g_1, g_2, \cdots, g_m\}$ and g_{m+1} indicates the base station (BS), $n > m$.
- R_S denotes the maximum communication range of the sensor nodes.

- $dis(s_i, g_j)$ denotes the Euclidian distance between sensor node s_i and gateway g_j.
- d_i denotes the traffic load contributed by a sensor node s_i, $s_i \in S, d_i \in \mathbb{Q}$ where $\mathbb{Q}$ is the set of rational numbers. It may be noted that the sensor nodes may have practically different processing and communication capabilities for WSNs that are heterogeneous in nature. Therefore, the traffic load contributed by the sensor nodes may vary depending on rate of data generation and communication. It is assumed that the traffic load contributed by each sensor node is estimated prior to the formation of clusters, as assumed in [32].
- G_i denotes the set of gateways, which are within communication range (R_S) of sensor node s_i. In other words,

$$G_i = \{g_j | dis(s_i, g_j) \leq R_S \wedge g_j \in \xi\} \tag{4.1}$$

Therefore, s_i can be assigned to any one of the gateways from G_i, where $G_i \subseteq \xi$. For example, $G_r = \{g_1, g_3, g_7\}$ means that s_r can be assigned to any one of the CHs, g_1, g_3, g_7.

- Let W_j be the load assigned to the cluster head g_j. In other words,

$$W_j = \sum_{i=1}^{n} \{d_i \times \alpha_{ij}\} \tag{4.2}$$

where α_{ij} is a Boolean variable such that

$$\alpha_{ij} = \begin{cases} 1 & \text{If sensor node } s_i \text{ is assigned to cluster head } g_j, \\ 0 & \text{otherwise} \end{cases} \tag{4.3}$$

Then the overall maximum load of cluster heads is $W = \max\{W_i | \forall g_i \in \xi\}$. Now, let us address the problem of clustering, where the main objective is to minimize the overall maximum load of the CHs. Then the linear programming (LP) of the load-balanced clustering problem can be formulated as follows:

$$\text{Minimize } W = \max\{W_i | \forall g_i \in \xi\}$$

Subject to

$$\sum_{j=1}^{m} \alpha_{ij} = 1 | \forall s_i \in S \tag{4.4}$$

$$\sum_{s_i \in S} d_i \times \alpha_{ij} \leqslant W | \forall g_j \in G_i \tag{4.5}$$

$$\sum_{j=1}^{m} dis(s_i, g_j) \times \alpha_{ij} \leqslant R_S | \forall s_i \in S \tag{4.6}$$

The constraint (4.4) states that a sensor node can be assigned to one and only one gateway and (4.5) indicates that the total load of all the sensor nodes assigned to a gateway must not exceed the overall maximum load of the gateway. The constraint (4.6) ensures that the sensor nodes are assigned to the gateway within their communication range.

4.3 Terminologies

The following terminologies are used in the algorithms. Depending on the communication range between the sensor nodes and the gateways, two kinds of sensor nodes are defined in the system: the *restricted node*, and *open node* as follows [136, 137]:

1. ***Restricted node and restricted set*:** Restricted nodes are those sensor nodes that can communicate with one and only one gateway. Restricted set is the set of all restricted nodes in the WSN. We refer this set as R_{set}. It is obvious to note that a sensor node s_i belongs to R_{set} if it satisfies the following criteria:

$$s_i \in R_{set} \Leftrightarrow [\{\exists g_j | g_j \in G_i \wedge g_j \in \xi\} \wedge \{\exists \neg g_k | g_k \in G_i \wedge g_k \in \{\xi - g_j\}\}] \quad (4.7)$$

   where G_i is the set of all those gateways that are within communication range of s_i and ξ is the set of all gateways as mentioned above.

2. ***Open node and open set*:** Open nodes are those sensor nodes that can communicate with more than one gateway. Open set is the collection of all open nodes in the WSN. We refer this set as O_{set}. A sensor node s_i belongs to O_{set} if it satisfies the following criteria:

$$s_i \in O_{set} \Leftrightarrow [s_i \notin R_{set}] \quad (4.8)$$

3. Assigning a sensor node to a CH depends on various factors such as its distance from the CH, its rate of data generation and communication to the CH, and its residual energy. We use $P_i(g_x)$ to denote the probability of assigning s_i to the cluster head g_x. Therefore, we have

$$\sum_{x=1}^{m} P_i(g_x) = 1 | \forall s_i \in S, g_x \in \xi \quad (4.9)$$

4. The expected load (*EL*) of a cluster head g_x is defined as the summation of mean loads of all the sensor nodes with the probability that they can be assigned to g_x. Therefore, the *EL* can be expressed as follows:

$$EL(g_x) = \sum_{i=1}^{n} P_i(g_x) \times d_i | \forall s_i \in S, g_x \in \xi \quad (4.10)$$

The EL is illustrated with the following example. Let us assume that there are three sensor nodes, s_1, s_2, and s_3, and their set of possible CHs are $G_1 = \{g_1, g_4, g_5\}, G_2 = \{g_2, g_4, g_7, g_{10}\}$, and $G_3 = \{g_2, g_4\}$. Here, each of G_1, G_2 and G_3 has g_4 as a common member. Let g_4 be not a member of any other set of CHs except G_1, G_2, and G_3. Therefore, $P_1(g_4) = \frac{1}{3}$. Similarly, $P_2(g_4) = \frac{1}{4}$ and $P_3(g_4) = \frac{1}{2}$. Then, using equation (4.10), we have

$$EL(g_4) = \frac{1}{3} \times d_1 + \frac{1}{4} \times d_2 + \frac{1}{2} \times d_3$$

If any sensor node, say s_i, is finally assigned to the CH, say g_x, then the EL of the cluster head gx and the EL of any other CH, say g_y can be updated as follows:

$$EL(g_x) = EL(g_x) + (1 - P_i(x)) \times d_i \tag{4.11}$$

$$EL(g_y) = EL(g_y) - P_i(y) \times d_i, \forall g_y \in G_i - \{g_x\} \tag{4.12}$$

5. Maximum possible load (MPL) of a cluster head g_x is the summation of the loads contributed by all the sensor nodes prior their allotment to g_x. It can be noted that g_x must be within the range of all such sensor nodes. Therefore the MPL of a cluster head g_x can be expressed as follows:

$$MPL(g_x) = \sum_{i=1}^{n} d_i | \forall s_i \in S \wedge g_x \in G_i \tag{4.13}$$

6. The current load (CL) of a cluster head g_x is the summation of the loads contributed by all sensor nodes after their allotment to g_x.

4.4 Load balanced clustering algorithms

Network setup is performed in two phases: bootstrapping and clustering. During the boot-strapping process, all the sensor nodes and gateways are assigned unique IDs. Then the sensor nodes broadcast their IDs using CSMA/CA MAC layer protocol. Therefore, the gateways within the communication range of these sensor nodes can collect the sensor IDs and finally send the local network information to the base station. Thus, for each sensor node, the number of gateways within its communication range can be calculated by the base station. In the clustering phase, base station executes the clustering algorithm. When the clustering is over, all the sensor nodes are informed about the ID of the gateway they belong to.

4.4.1 Algorithm for equal load

Here, it is considered to have each sensor node equal traffic load, i.e., $d_i = \alpha(say), \forall s_i \in S, 1 \leq i \leq n$. Therefore, minimizing the overall maximum load of each CH is equivalent to minimizing the maximum number of sensor nodes that can be assigned to each CH. The basic idea of our algorithm is as follows. We first sort the set of sensor nodes S in non-decreasing order on $|G_i|$ of $s_i, \forall s_i \in S, 1 \leq i \leq n$. Let $S = \{s_a, s_b, s_c, \cdots, s_p\}$ be this sorted sensor list. We now successively consider each sensor node from this list (starting with s_a) to assign it to the correct CH. In order to assign s_a, we consult its corresponding set of possible CHs, i.e., G_a, and calculate the EL values using equation (4.10) for all the CHs belonging to G_a. The sensor node s_a is assigned to that CH, which has the minimum EL value. In other words, s_a selects the cluster head, say g_x, only if

$$EL(g_x) = Min\{EL(g_k)|\forall g_k \in G_a\} \tag{4.14}$$

If there are two or more CHs with the same EL value, then select that CH for which the probability of assigning the sensor node is maximum. If the probability also ties, then select the CH that has the minimum number of sensor nodes already assigned to it. After each assignment of sensor node, the EL value of the CHs are updated by equations (4.11) and (4.12) for the assignment of the next sensor from the sorted list. The same procedure is continued until all the sensor nodes are allotted to their correct CH. The pseudo-code is formally presented in Algorithm 1.

Algorithm 1 2-Approx-Load-Balanced-Clustering (2ALBC)

Input: (1)Set of sensors $S = \{s_1, s_2, ..., s_n\}$ with traffic load $d_i \forall i, 1 \leq i \leq n$.
(2) Set of gateways $\xi = \{g_1, g_2, ..., g_m\}$, where $m < n$.
(3) $G_i, \forall s_i \in S$ and $P_i(x)[\equiv P_i(g_x)], \forall g_x \in G_i$.
Output: An assignment $A : S \rightarrow \xi$ such that the overall maximum traffic loads of CHs is minimized.

Sort the sensor nodes in non-decreasing order on the number of possible CHs upon which they can be assigned.
for $j = 1$ to n **do**
    $EL[k] = EL[k] + p_j(k) \times d_j, \forall g_k \in G_j$ /*$EL[k] \equiv EL[g_k]$*/
end for
for $j = 1$ to n **do** /* Assignment of sensor nodes to CHs */
    Select the CH (say g_k) with minimum EL value using equation (4.10) and assign s_j to g_k. If there are more than two CHs with the same EL value, then select that CH with highest probability to join. If the probability is also the same, then select the CH that has the minimum number of sensor nodes already assigned to it.
    $EL[k] = EL[k] + (1 - p_j(k)) \times d_j$
    $EL[x] = EL[x] - p_j(x) \times d_j, \forall g_x \in \{G_j - g_k\}$
end for
Stop.

TABLE 4.1: Sensor nodes with the list of possible gateways

Sensor nodes (s_i)	Possible gateways (G_i)
s_1	$G_1 = \{g_1, g_2, g_3, g_4\}$
s_2	$G_2 = \{g_1, g_2\}$
s_3	$G_3 = \{g_2, g_3\}$
s_4	$G_2 = \{g_1, g_2, g_3, g_4\}$
s_5	$G_5 = \{g_2, g_3, g_4\}$
s_6	$G_6 = \{g_1, g_2, g_3, g_4\}$
s_7	$G_7 = \{g_1, g_2, g_3\}$
s_8	$G_8 = \{g_1, g_2, g_4\}$
s_9	$G_9 = \{g_1, g_3\}$
s_{10}	$G_{10} = \{g_2, g_3, g_4\}$
s_{11}	$G_{11} = \{g_1, g_2, g_3, g_4\}$
s_{12}	$G_{12} = \{g_1, g_2, g_3, g_4\}$
s_{13}	$G_{13} = \{g_2, g_3\}$
s_{14}	$G_{14} = \{g_1, g_2\}$
s_{15}	$G_{15} = \{g_1, g_3, g_4\}$

4.4.2 An illustration

Consider a WSN of 15 sensor nodes and 4 gateways, i.e., $S = \{s_1, s_2, s_3, \cdots, s_{15}\}$ and $\xi = \{g_1, g_2, g_3, g_4\}$. Without loss of generality, let us assume that $d_i = 1$. Table 4.1 shows the possible gateways to which the sensor nodes may be assigned. In our algorithm, sensor nodes are sorted in nondecreasing order on G_i. The sorted list of the sensor nodes is shown in Table 4.2.

Initially, every CH has no load. As seen from the Table 4.2, s_2 is the first sensor node that can be assigned to any one of the cluster heads g_1 and g_2. Therefore, we calculate the EL values of g_1 and g_2 as follows: It is observed from Table 4.2 that the possible sensor nodes that may be assigned to g_1 are $s_1, s_2, s_4, s_6, s_7, s_8, s_9, s_{11}, s_{12}, s_{14}$, and s_{15}. It is also clear from Table 4.1 that s_1 has four possible CHs. So, the probability of assigning s_1 to g_1 is $\frac{1}{4}$, i.e., $P_1(g_1) = \frac{1}{4}$. Similarly, the probabilities of assigning $s_2, s_4, s_6, s_7, s_8, s_9, s_{11}, s_{12}, s_{14}$, and s_{15} to g_1 are $\frac{1}{2}, \frac{1}{4}, \frac{1}{4}, \frac{1}{3}, \frac{1}{3}, \frac{1}{2}, \frac{1}{4}, \frac{1}{4}, \frac{1}{2}$, and $\frac{1}{3}$ respectively. Therefore, by equation (4.10) we obtain

$$
\begin{aligned}
EL(g_1) &= \frac{1}{4}.1 + \frac{1}{2}.1 + \frac{1}{4}.1 + \frac{1}{4}.1 + \frac{1}{3}.1 + \frac{1}{3}.1 \\
&\quad + \frac{1}{2}.1 + \frac{1}{4}.1 + \frac{1}{4}.1 + \frac{1}{2}.1 + \frac{1}{3}.1[\because d_i = 1] \\
&= 3.75
\end{aligned}
$$

Similarly, we calculate $EL(g_2) = 4.583$. As g_1 has lesser EL value than g_2, we

TABLE 4.2: Sorted list of sensor nodes with possible gateways

Sensor nodes (s_i)	Possible gateways (G_i)
s_2	$G_2 = \{g_1, g_2\}$
s_3	$G_3 = \{g_2, g_3\}$
s_9	$G_9 = \{g_1, g_3\}$
s_{13}	$G_{13} = \{g_2, g_3\}$
s_{14}	$G_{14} = \{g_1, g_2\}$
s_5	$G_5 = \{g_2, g_3, g_4\}$
s_7	$G_7 = \{g_1, g_2, g_3\}$
s_8	$G_8 = \{g_1, g_2, g_4\}$
s_{10}	$G_{10} = \{g_2, g_3, g_4\}$
s_{15}	$G_{15} = \{g_1, g_3, g_4\}$
s_1	$G_1 = \{g_1, g_2, g_3, g_4\}$
s_4	$G_4 = \{g_1, g_2, g_3, g_4\}$
s_6	$G_6 = \{g_1, g_2, g_3, g_4\}$
s_{11}	$G_{11} = \{g_1, g_2, g_3, g_4\}$
s_{12}	$G_{12} = \{g_1, g_2, g_3, g_4\}$

assign s_2 to g_1. After assigning, $EL(g_1)$ is updated by equation (4.11) as follows:

$$\begin{aligned} EL(g_1) &= EL(g_1) + (1 - P_2(g_1)).1 \\ &= 3.75 + 0.5[\because P_2(g_1) = \frac{1}{2}] \\ &= 4.25 \end{aligned}$$

The EL value of g_2 is also updated by equation (4.12) as follows:

$$\begin{aligned} EL(g_2) &= EL(g_2) - P_2(g_1) \\ &= 4.583 - 0.5[\because P_2(g_2) = \frac{1}{2}] \\ &= 4.083 \end{aligned}$$

Next we consider s_3 for its assignment. From Table 4.1, it can be noted that s_3 may be assigned to g_2 or g_3. Now we calculate the EL value of g_3 as follows:

$$\begin{aligned} EL(g_3) &= \frac{1}{4}.1 + \frac{1}{2}.1 + \frac{1}{4}.1 + \frac{1}{3}.1 + \frac{1}{4}.1 + \frac{1}{3}.1 + \frac{1}{2}.1 \\ &\quad + \frac{1}{3}.1 + \frac{1}{4}.1 + \frac{1}{4}.1 + \frac{1}{2}.1 + \frac{1}{3}.1 \\ &= 4.083 \end{aligned}$$

As the EL values of g_2 and g_3 are the same, $P_3(g_2)$ and $P_3(g_3)$ are also the same and both g_2 and g_3 are zero loaded, and s_3 can be assigned to any one of g_2 and g_3. We assign it to g_2. Now we update EL values of g_2 and g_3 and continue the same method for assigning the remaining sensor nodes. The successive assignment of the

TABLE 4.3: Assignment of successive sensor nodes to the CHs

Sensors with possible cluster heads	Cluster head (g_i)			
	g_1	g_2	g_3	g_4
	Assigned sensors and *EL*	Assigned sensors and *EL*	Assigned sensors and *EL*	Assigned sensors and *EL*
s_2 $\{g_1, g_2\}$	s_2, EL=4.25	EL=4.083	EL=4.083	EL=2.583
s_3 $\{g_2, g_3\}$	s_2, EL=4.25	s_3, EL=4.583	EL=3.583	EL=2.583
s_9 $\{g_1, g_3\}$	s_2, EL=3.75	s_3, EL=4.583	s_9, EL=4.083	EL=2.583
s_{13} $\{g_2, g_3\}$	s_2, EL=3.75	s_3, EL=4.083	s_{13}, s_9, EL=4.583	EL=2.583
s_{14} $\{g_1, g_2\}$	s_2, s_{14}, EL=4.25	s_3, EL=3.583	s_{13}, s_9, EL=4.583	EL=2.583
⋮	⋮	⋮	⋮	⋮
s_{12} $\{g_1, g_2, g_3, g_4\}$	s_{14}, s_2, s_{15}, s_6, EL=4.0	s_3, s_7, s_{10}, s_{11}, EL=4.0	s_{13}, s_9, s_1, s_{12}, EL=4.0	s_5, s_8, s_4, EL=3.0

sensor nodes s_3, s_9, and finally s_{12} to their CHs along with their calculated EL values are shown in Table 4.3.

Lemma 4.1. *The above* 2-*Approx-Load-Balanced-Clustering algorithm* (2*ALBC*) *produces the optimal solution for equal load of the sensor nodes.*

Proof. First, note that the sensor list S is initially sorted in non-decreasing order on the number of CHs. Let $S = \{s_1, s_2, s_3, \ldots, s_n\}$ be this sorted list. Now, the algorithm assigns $s_1, s_2, s_3, s_4, \ldots, s_n$ successively. For the assignment, the following approach is followed for load balancing. The algorithm first assigns that sensor node which has the least chance of being assigned to a CH. As a result, any other sensor node having more chance can select a CH with the least load.

Let s_k be the last of a assigned sensor node to the maximum loaded cluster head g_r after its assignment. Let current load on g_r be l_r. Suppose there is a cluster head $g_s \in G_k$ with total load l_s such that $l_s < l_r$. Therefore, s_k can be assigned to g_s and the current load of g_s will be $l_s + \alpha$, where α is the load of each sensor node. Then the algorithm is not optimal if $(l_s + \alpha) < l_r$. But at the time of assignment of s_k, the selected cluster head g_r had the minimum EL value, i.e., $(l_r - \alpha)$ was minimum. So, after assigning of s_k to any $g_s \in G_k - \{g_r\}$, $(l_s + \alpha) \geq l_r$. Hence the proof. □

Lemma 4.2. *The time complexity of the algorithm* 2*ALBC is* $O(n \log n)$.

Proof. Line 1 of the Algorithm 1 requires $O(n \log n)$ time for sorting n sensor nodes. In the for loop of the line 2, a CH list is created where the average possible load of each CH is calculated. In the worst case, the loop can iterate m times. So, line 2-4 can take $O(mn)$ time in worst case. The for loop in line 5 iterates n times in which line 6 and line 8 are the dominating steps requiring $O(m)$ time for each in the worst situation. Therefore, the loop can be executed in $O(mn)$ time. Thus the above algorithm requires $O(mn)$ time. If $m < \log n$, it requires $O(n \log n)$ time. □

Theorem 1. *The algorithm 2ALBC produces an optimal solution in $O(n \log n)$ time assuming equal loads of the sensor nodes.*

Proof. Follows from Lemma 4.1 and 4.2. □

4.4.3 Approximation algorithms for unequal load

Here, a WSN scenario is considered in which the traffic load contributed by the sensor nodes is variable. This is due to the fact that the sensor nodes may generate or process the data at different rates and their rate of communication with their CHs is also different.

Lemma 4.3. *Load balanced clustering problem with unequal load for sensor nodes is NP-hard.*

Proof. Please refer to [32]. □

Lemma 4.4. *The algorithm 2ALBC is a 2-approximation algorithm for load balanced clustering problem with unequal load of the sensor nodes.*

Proof. Let OPT be the maximum load of a CH in an optimal solution. Then it is obvious to note that $OPT \geq d_i, \forall i, 1 \leq i \leq n$. Let I be the smallest instance of the problem for which the algorithm produces results not as good as the optimal solution. Let g_i be a CH with maximum load after a complete run of the load-balanced clustering algorithm, i.e., $W_i = \max \{W_j | \forall j, 1 \leq j \leq n\}$. Let s_r be the last sensor node assigned to g_i. The crucial property of our algorithm is that, at the time of the assignment of s_r, g_i was the minimum loaded CH from G_r. So, before assignment of s_r, the load of g_i was $(W_i - d_r)$, which is less than or equal to $OPT(I)$. Therefore, $W_i - d_r \leq OPT(I)$, i.e., $W_i \leq OPT(I) + OPT(I)$ [as, $d_r \leq OPT(I)$]. Hence, $W_i \leq 2OPT(I)$. □

Here, it is assumed that the traffic loads contributed by the sensor nodes are unequal and already calculated prior to cluster formation. The basic idea is as follows. First, all the sensor nodes are assigned to their corresponding gateway. Then all the sensor nodes are sorted in non-increasing order on their contributing traffic load. Now, the basic principle of the algorithm is to assign a sensor node that contributes maximum load to a minimum loaded CH within its range. After assignment, current load and maximum possible load of the CHs are updated. Then the sensor node contributing the next maximum load is assigned to the current minimum loaded CH within its range. This process is continued until all the sensors are assigned to the CHs. The algorithm is now formally presented in Algorithm 2.

Lemma 4.5. *The time complexity of* 1.5*-Approx-Load-Balanced-Clustering algorithm* (1.5*ALBC*) *is $O(n \log n)$.*

Proof. The while loop of line 1 requires $O(m)$ time in worst case to assign the R_{set} to their corresponding CH. In the for loop of line 5, a CH list is created where the maximum possible load of each CH is calculated. In the worst case, it can take $O(mn)$

Algorithm 2 1.5-Approx-Load-Balanced-Clustering (1.5ALBC)

Input: (1)Set of sensors $S = \{s_1, s_2, ..., s_n\}$ with traffic load $d_i \forall i, 1 \leq i \leq n$.
(2) Set of gateways $\xi = \{g_1, g_2, ..., g_m\}$, where $m < n$.
(3) For each s_i, the set of cluster heads G_i.
Output: An assignment $A : S \rightarrow \xi$ such that the overall maximum traffic loads of CHs is minimized.

1: **while** ($R_{set} \neq NULL$) **do**
2:     Assign successive sensor node s_i to their corresponding cluster head (say g_k) and delete s_i from R_{set}.
3:     $CL[k] = CL[k] + d_i$
4: **end while**
5: **for** $j = 1$ to m **do**
6:     $MPL[j] = CL[j] + \sum(d_i | g_j \in G_i \bigwedge s_i \in O_{set})$
7: **end for**
8: Sort the sensor nodes from O_{set} in non-decreasing order on the number of possible CHs upon which they can be assigned.
9: **while** $O_{set} \neq NULL$ **do**
10:     Select successive sensor node (say s_i) from O_{set}.
11:     Select the cluster head (say g_k) from G_i with minimum current loaded (CL). If there are more than two CHs with the same current load (CL), then select the CH that has minimum MPL value.
12:     Assign s_i to g_k and delete s_i from g_k.
13:     $CL[k] = CL[k] + d_i$
14:     $MPL[x] = MPL[x] - d_i, \forall g_x \in \{G_i - g_k\}$
15: **end while**
16: Stop.

---

time. The sorting in line 8 requires $O(n \log n)$ time in worst case for sorting sensor nodes of O_{set}. In the while loop of line 9, a sensor node is assigned to a CH. So it can be executed in $O(mn)$ time. Therefore the above algorithm requires $O(mn)$ time. If $m < \log n$, it requires $O(n \log n)$ time. □

Lemma 4.6. *The* 1.5*ALBC is a* 1.5*-Approximation algorithm of load balanced clustering problem* (*LBCP*) *with unequal loads of the sensor nodes.*

Proof. Let OPT be the maximum load of a CH in an optimal solution. It is obvious to note that $OPT \geq d_i, \forall i, 1 \leq i \leq n$. Let I be the smallest instance of the problem for which this algorithm conflicts with the optimal solution. Let g_i be a CH with maximum load after a complete run of the algorithm 1.5*ALBC*, i.e., $W_i = \max \{W_j | \forall j, 1 \leq j \leq m\}$.

Let s_r be the last sensor node assigned to g_i. Therefore, s_r is the sensor node that has the minimum load between all sensor nodes. Now, at the time of the assignment of s_r, g_i was the minimum loaded CH from G_r. As we are assuming that I is the smallest instance, the algorithm conflicts with the optimal solution only after assignment of s_r. Without loss of generality, we may assume that each CH can be assigned with at least two sensor nodes. Therefore, all the CHs except g_i are assigned by at least two sensor nodes. This implies that $d_r \leq OPT(I)/2$, as s_r is the least loaded sensor node. Therefore, before allotment of s_r, load of g_i was $W_i - d_r$, which is less than or equal to $OPT(I)$. In other words, $W_i - d_r \leq OPT(I)$ i.e., $W_i \leq OPT(I) + OPT(I)/2$ [as, $d_r \leq OPT(I)$]. Hence, $W_i \leq 1.5OPT(I)$. □

4.4.4 Performances based on load balancing

To judge the quality of the load balancing, the standard deviation of the gateway load is measured and plotted against the number of sensor nodes. The standard deviation (σ) of the gateway load gives even distribution of the load per gateway. If there are m gateways and n sensor nodes, the standard deviation of gateway load is given by

$$\sigma = \sqrt{\frac{1}{m}\sum_{j=1}^{m}(\mu - W_j)^2} \tag{4.15}$$

where μ(average load) $= \frac{1}{m}\sum_{i=1}^{n} d_i$, d_i is the load of the sensor node s_i and W_j is the overall load of the gateway g_j.

The approximation algorithms are compared with various well known algorithms including GLBCA [32], LBC [38] and LDC [33]. The comparison results in terms of load balancing for unequal load of the sensor nodes are shown in Figure 4.1(a). The approximation algorithms perform better in load balancing. LBC and LDC perform poorly as they have not considered unequal load of the sensor nodes.

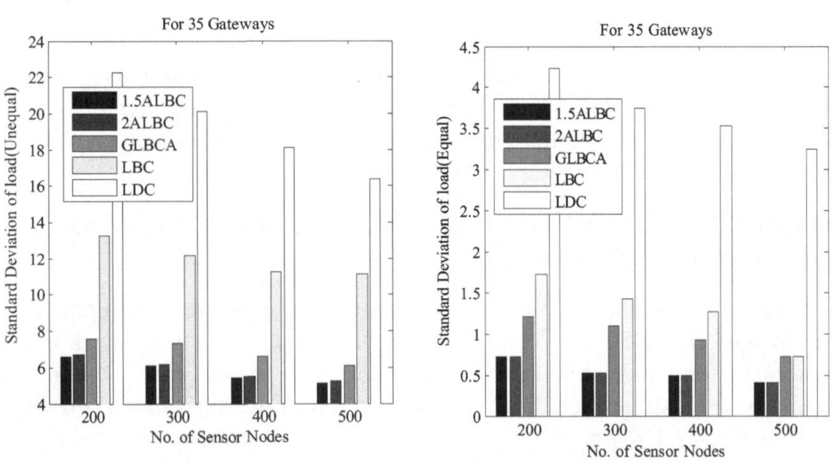

FIGURE 4.1: Comparison of load balancing for (a) unequal load and (b) equal load of the sensor nodes using 35 gateways

The algorithms are also compared for equal load of the sensor nodes, the comparison results of which are depicted in Figure 4.1(b). This can be noted that the 2ALBC and 1.5ALBC perform same as GLBCA, they perform far better than Genetic Algorithm (GA), LBC and LDC. The rationale behind is that all three algorithms 2ALBC, 1.5ALBC and GLBCA are known to be optimal for the equal load of the sensor nodes whereas the GA, LBC and LDC are known to be suboptimal in terms of load balancing. Note that LBC can be executed separately for load balancing and energy consumption issues but not considering both of them together. In the simulation, LBC is run by considering load balancing of the CHs only.

4.4.5 Execution time

Execution time of the algorithms are also measured. It can be observed from Figure 4.2(a)-(b) that the 1.5ALBC and 2ALBC are better than LBC and far better than GLBCA in terms of execution time. The rationale behind this is that in order to assign a sensor node, GLBCA builds a bipartite graph of the sensor nodes and the CHs to find out the maximum matching for assigning a sensor node to a CH. The time complexity of the algorithm is $O(n[n + m + q])$ in worst case (m number of gateways and n number of sensor nodes and q is the cardinality of the bipartite graph). Thus GLBCA takes considerably high time for large number of sensor nodes. LBC takes time to expands R_{Set} based on the distance from the sensor node to gateway and in worst case this process takes $O(mn)$ time. Then, LBC assigns the R_{Set} which also takes $O(mn)$ time and finally assign all remaining sensor nodes which takes $O(mn \log n)$ time. On the other hand 2ALBC and 1.5ALBC use only sorting and the overall time complexity of these algorithms is $O(n \log n)$ in worst case. Thus 2ALBC

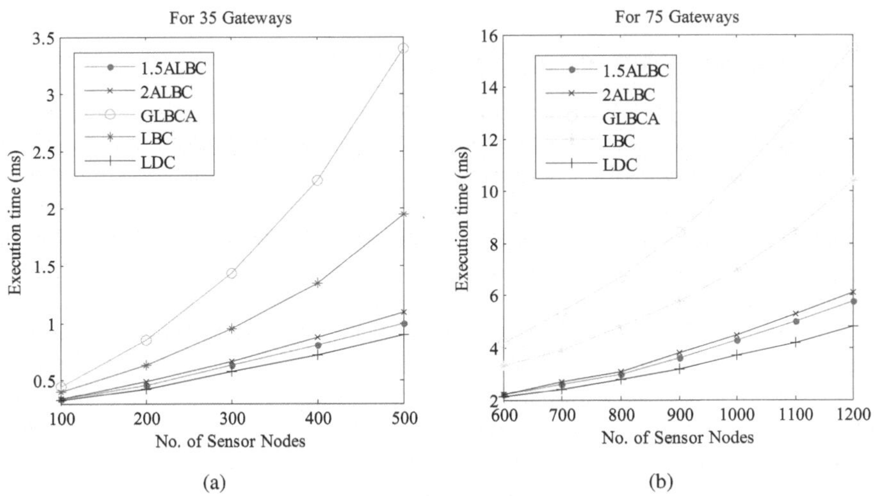

FIGURE 4.2: Comparison of execution time using (a) 15 gateways and (b) 30 gateways

Algorithm 3 Heap-based-Load-Balanced-Clustering-Algorithm(HLBCA)

Input: (1) A set of sensor nodes $S = \{s_1, s_2, ..., s_n\}$ with traffic load $d_i, \forall i, 1 \leq i \leq n$.
(2) The set of cluster heads $\xi = \{g_1, g_2, \cdots, g_m\}$ and $G_i, \forall s_i \in S$.

Output: An assignment $A : S \rightarrow \xi$ such that the overall maximum traffic loads of CHs is minimized.

$Build\_Max\_Heap(H, S, n)$. /*Build a max heap H of $S = \{s_1, s_2, ..., s_n\}$ on $d_i, \forall s_i \in S$*/
for $j = 1$ to m
    $CL[j] = 0$ /* initially, current load of gateways is zero.*/
end for
for $j = 1$ to m
    Select the rooted sensor node (say s_k).
    Assign s_k to the minimum loaded gateway (say g_j) amongst G_k.
    Replace $H(1)$ by $H(n)$. /*$H(i)$ be the ith element of the heap.*/
    Delete $H(n)$.
    $n = n - 1$.
    $Max\_Heapify(H, 1, n)$
    $CL[j] = CL[j] + d_k$ /*After assigning s_k to g_j*/
end for
Stop.

and 1.5ALBC consume very less execution time than GLBCA and LBC. It can be observed that, LDC executes faster than all the algorithms. This is because LDC simply assigns the sensor nodes to its nearest CH and thus it takes less execution time.

4.4.6 Heap based approximation algorithm

The basic principle of the algorithm is to assign first the sensor node, which contributes maximum load to a minimum loaded gateway within its communication range. First, a max-heap of the sensor nodes is built depending on their contributing traffic load . Then we pick up the rooted sensor node from the heap, i.e., the node with maximum traffic load and assign it to the minimum loaded gateway within its range. Now, the rooted node from the heap, i.e., the sensor node which is already assigned, is deleted and replaced by the last element of the heap. Thus the heap size is decreased by one through deleting the assigned sensor node. Now, the heap is max-heapified (Cormen, T. H. et al., 2009)[128] at the rooted position so that the next maximum loaded sensor node is once again at the root. Again we pick up the rooted sensor node from the heap and assign it into a minimum loaded gateway and then follow the same. The same procedure is continued until all the sensor nodes are allotted to their correct gateway. The pseudo code of the approach is formally presented in Algorithm 3.

Lemma 4.7. *The time complexity of the algorithm HLBCA is $O(nm)$.*

Proof. Step 1 requires $O(n)$ time (Cormen, T. H. et al., 2009)[128] to build a max heap of n sensor nodes. The for loop of the Step 2 requires $O(m)$ time. Step 3 iterates exactly n times for n number of sensor nodes in which Step 3.1 requires $O(m)$ times in worst case, Step 3.2 to step 3.4 can be executed in constant time, Step 3.5 takes $O(\log n)$ times and Step 3.6 also requires constant time. Therefore, total time taken by Step 3 is,

$$n \times \{O(m) + O(1) + O(1) + O(1) + O(\log n) + O(1)\}$$
$$= O(mn) + O(n\log n) + O(n)$$

However, the time for $Max\_Heapify(H, 1, n)$ in Step 3.5 varies on the height of the node for which it is called. The time required by $Max\_Heapify(H, 1, n)$ when it is called for a node of height h is $O(h)$. It is noteworthy that, for n number of node, the height of the tree (heap) is $\lfloor \log n \rfloor$, assuming that the root is at the maximum height. It can be observed that, the maximum number of nodes at the height h of a binary tree of n nodes is $2^{\lfloor \log n \rfloor - h}$. Therefore, total time required by Step 3.5 for n number of nodes can be expressed as follows:

$$\sum_{h=0}^{\lfloor \log n \rfloor} h \times 2^{\lfloor \log n \rfloor - h} = \sum_{h=0}^{\lfloor \log n \rfloor} h \times \frac{2^{\lfloor \log n \rfloor}}{2^h} \leq \sum_{h=0}^{\lfloor \log n \rfloor} h \times \frac{n}{2^h}$$
$$= n \sum_{h=0}^{\lfloor \log n \rfloor} \frac{h}{2^h} \leq n \sum_{h=0}^{\infty} \frac{h}{2^h} = 2n \left[\because \sum_{h=0}^{\infty} \frac{h}{2^h} = 2 \right]$$
$$= O(n)$$

Therefore, Step 3 can be executed in $O(mn) + O(n) + O(n)$ time, i.e., in $O(mn)$ time. As Step 3 dominates the Step 1 and Step 2, the above algorithm requires $O(mn)$ time in worst case. □

Lemma 4.8. *The above Heap-based-Load-Balanced-Clustering algorithm (HLBCA) produces optimal solution for equal load of the sensor nodes.*

Proof. The basic principle of the algorithm is to assign first that sensor node, which contributes maximum load to a minimum loaded gateway. However, if we assume that each sensor node has equal traffic load, i.e., $d_i = \alpha$ (say), $\forall s_i \in S, 1 \leq i \leq n$, then the order of the assignment of the sensor nodes will not affect in the result in terms of load balancing. Therefore, in each assignment of the sensor nodes the algorithm assigns a sensor node to the minimum loaded gateway to increase its load.

Let us assume that s_k be the last assigned sensor node which is assigned to the gateway g_r and after the assignment, g_r is the maximum loaded gateway. Let current load on g_r be W_r. Therefore, for the gateways $g_i, \forall g_i \in \{G_k - g_r\}, W_i \leq W_r$. If s_k can be assigned to any one of $g_i, \forall g_i \in \{G_k - g_r\}$, then the load of g_i would be, $W_i + d_k$, i.e., $W_i + \alpha$, where α is the load of each sensor node. Then the algorithm is not optimal if $(W_i + \alpha) \leq W_r$. But it should be noteworthy that at the time of assignment of s_k, the selected gateway g_r had the minimum load, i.e., $(W_r - \alpha)$ was minimum. Therefore, if s_k was assigned to any $g_i, \forall g_i \in \{G_k - g_r\}$ instead of g_r, then it would be $(W_i + \alpha) \geq W_r$. Hence proof. □

Theorem 2. *The algorithm HLBCA produces optimal solution in O(nm) time assuming equal loads of the sensor nodes.*

Proof. Follows from Lemma 4.7 and Lemma 4.8. □

Lemma 4.9. *The HLBCA is a 1.5-Approximation algorithm of Load Balanced Clustering Problem with unequal loads of the sensor nodes.*

Proof. Let OPT be the maximum load of a gateway in an optimal solution. It is obvious to note that $OPT \geq d_i, \forall i, 1 \leq i \leq n$. Let I be the smallest instance of the problem for which this algorithm conflicts with optimal solution. Let g_i be a CH with maximum load after complete run of the algorithm HLBCA, i.e., $W_i = max\{W_j \mid \forall i, 1 \leq i \leq n\}$.

Let s_r be the last sensor node assigned to g_i. Therefore, sr is the sensor node which has the minimum load between all sensor nodes. Now at the time of the assignment of s_r, g_i was the minimum loaded CH from G_r. As we are assuming that I is the smallest instance, so the algorithm conflicts with optimal solution only after assignment of s_r. Without loss of generality, we may assume that each gateway has at least two sensor nodes those are allotted to it. Therefore all the gateways except g_i are being assigned by at least two sensor nodes. This implies that $d_r \leq OPT(I)/2$, as s_r is the least loaded sensor node. Therefore, before allotment of s_r, load of g_i was $(W_i - d_r)$, which is less than or equal to $OPT(I)$. In other words, $(W_i - d_r) \leq OPT(I)$, i.e., $(W_i \leq (OPT(I) + OPT(I)/2)$ as $d_r \leq OPT(I)/2$. Hence, $W_i \leq 1.5OPT(I)$. □

Theorem 3. *The algorithm HLBCA is an 1.5-Apprimation algorithm with polynomial time O(nm).*

Proof. Follows from Lemma 4.7 and Lemma 4.9. □

It is observed that the algorithm HLBCA is better than 2ALBC and LBC and far better than GLBCA in terms of execution time but slightly poorer than EELBCA[140]. The rationale behind this is that HLBCA uses heap of sensor nodes to efficiently balances the load of the CHs and it only iterates the function max_heapify which has a less execution time $O(logn)$ [128]. The overall time complexity of HLBCA is $O(mn)$. Thus HLBCA consumes very less execution time than GLBCA, LBC and 2ALBC.

4.5 Chapter summary

In this chapter, three load balanced clustering algorithms are described for wireless sensor networks. The first algorithm has been shown to be optimal in assigning sensor nodes to the CHs in the case of equal load of the sensor nodes. The algorithm has been shown to run in $O(n \log n)$ time. It is shown that the same scheme can also work as a 2-approximation algorithm for the situation where the loads of the sensor nodes can vary. Therefore, it is an improvement over the algorithm as proposed by Low et al [32] which requires $O(mn^2)$ time for equal load and $O(n[n + m + q])$ for unequal load of the sensor nodes.

Next, an improved algorithm have been presented that has 1.5-approximation ratio for the general case which is shown to run in $O(n \log n)$ time. Finally, a heap based load balanced clustering algorithms is presented for wireless sensor networks. The algorithm has been shown to be optimal in assigning sensor nodes to the CHs in the case of equal loads of the sensor nodes and is a 1.5-approximation algorithm for the general case, i.e., unequal load of the sensor nodes. The algorithm has been shown to run in $O(mn)$ time. It is observed that the algorithms outperform the existing algorithms in terms of load balancing for both of the equal and unequal load of the sensor nodes and execution time.

However,The clustering algorithms consider only load balancing issue with respect to data generation and communication of the member sensor nodes like the algorithms [32, 33, 38]. The residual energy of CHs and the communication distance is not considered. Both these issues are considered in other algorithm which are presented in the later chapters.

Chapter 5

Centralized Algorithms

Chapter 5

Centralized Algorithms

5.1 Introduction

Here, we describe some centralized algorithms. We first discuss a clustering algorithm called EELBCA (energy efficient load balanced clustering algorithm) is presented for WSNs. EELBCA bridges between both the issues, i.e., load balancing of the CHs and the energy saving of the sensor nodes. The algorithm is based on min-heap built with the CHs and shown to run in $O(n \log m)$ time for n number of sensor nodes and m CHs. This is an improvement over $O(mn \log n)$ and $O(mn^2)$ time algorithms as reported in [32] and [38] respectively.

In the EELBCA, load of the CHs is considered only in terms of the number of allotted sensor nodes to a CH. We do not consider the load of the CHs in terms of energy consumption along with communication overhead. Therefore, a parameter based clustering algorithm (PBCA) is also presented. PBCA balances the load of the CHs in terms of energy consumption of the CHs (incorporating load of communication to the base station) rather than the cardinality of the clusters, i.e., the number of sensor nodes assigned to a CH. It also considers the energy consumption of the sensor nodes. The first parameter used in PBCA controls the load balancing of the CHs and the second one controls the energy consumption of the sensor nodes. By suitably changing the value of the parameters, we can concentrate more on load balancing of the CHs or more on the energy consumption of the sensor nodes. Through experimental simulation, it is shown that both the algorithms outperform the similar works reported in [32] and [38], in terms of execution time, load balancing, energy consumption, number of live sensor nodes and number of live CHs.

5.2 Load balancing algorithms

Here, network setup is performed in two phases: bootstrapping and clustering. During the bootstrapping process, all the sensor nodes and gateways are assigned unique IDs. Then the sensor nodes broadcast their IDs using CSMA/CA MAC layer protocol. Therefore, the gateways within the communication range of these sensor nodes can collect the sensor IDs and finally send the local network information to the

Algorithm 4 Energy-Efficient-Load-Balanced-Clustering Algorithm (EELBCA)

Input: (1)Set of sensors $S = \{s_1, s_2, ..., s_n\}$.
(2) Set of gateways $\xi = \{g_1, g_2, ..., g_m\}$; where, $m < n$.
(3) $dis(s_i, g_j)$: All distance between s_i to g_j; $g_j \in G_i, R_{set}$ and O_{set}.
Output: An assignment $A : S \rightarrow \xi$ such that the overall maximum number of sensor nodes to CHs and total consumed energy is minimized.

while ($R_{set} \neq NULL$) **do**
  Assign successive sensor nodes s_i to their corresponding gateway g_j such that $s_i \in R_{set}$ and $g_j \in G_i$ and delete s_i from R_{set} and S.
end while
Build a min-heap of the gateways on the number of allotted sensor nodes to the gateways.
while ($S \neq NULL$) **do**
  Pick up the root node of the min-heap, say g_j
  Select and assign a sensor (Open) node si to g_j such that $g_j \in G_i$ and s_i is nearest sensor node to g_j.
  Delete s_i from S
  Adjust the min-heap so that the minimum loaded gateway will be at root
end while
Stop.

base station. Thus for each sensor node, the number of gateways within its communication range can be calculated by the base station. In clustering phase base station executes the clustering algorithm. When the clustering is over, all the gateways provide their IDs to their member sensor nodes by single hop communication. Then the gateways provide a TDMA schedule to their member sensor nodes for intra cluster communication.

It may be noted that the sensor nodes may have different processing and communication capabilities for WSNs which are heterogeneous in nature. Therefore, the traffic load contributed by the sensor nodes may vary depending on the data generation and communication rates. However, in this scenario a homogeneous network is assumed in which sensor nodes have equal processing and communication load. Therefore, the main objective is to balance the load of the gateways in terms of the number of allotted sensor nodes to the gateways.

At first the restricted nodes are assign to their corresponding gateway. Then a min-heap is built using the gateways depending on their respective number of assigned sensor (restricted) nodes. Let $\psi = \{g_1, g_2, g_3, \ldots, g_m\}$ be the list of gateways after building the min-heap. So, g_1 has been assigned by the minimum number of sensor nodes. Now we assign that s_i to g_1 which is the nearest to g_1 and $g_1 \in G_i$. Now, we rearrange the min-heap. So, the gateway with next minimum number of sensor nodes assigned to it is at the root of the min heap. We pick up this gateway and assign a sensor node from O_{set} which is nearest to this gateway and within its communication range. The same procedure is continued until all the sensor nodes

TABLE 5.1: Distance (meter) between sensor nodes to gateways

Gateways	Sensor nodes											
	s_1	s_2	s_3	s_4	s_5	s_6	s_7	s_8	s_9	s_{10}	s_{11}	s_{12}
g_1	∞	∞	26.3	39.5	38.4	∞	∞	∞	51.3	32.5	58.5	52.4
g_2	41.2	48.5	38.4	∞	33.6	42.8	57.9	52.1	∞	59.6	18.6	57.3
g_3	30.5	∞	52.6	∞	∞	31.7	43.7	∞	56.4	56.9	∞	17.5

are allotted to their correct gateway. The algorithm is formally given in Algorithm 4. The algorithm is illustrated by the following illustration [140, 141].

An illustration: Consider a WSN of 12 sensor nodes and 3 gateways, i.e., $S = \{s_1, s_2, \ldots, s_{12}\}$ and $\xi = \{g_1, g_2, g_3\}$. Table 5.1 shows the distances between the sensor nodes and the gateways. Here, we assume that the communication range of sensor nodes is 60 meters, i.e., $d_0 = 60$ meters.

The ∞ distance in Table 5.1 indicates that the sensor node is not within the communication range of the corresponding gateway. For example s_9 is not within communication range of g_2 and in the same way s_4 is not within communication range of g_2 and g_3. Depending on the number of possible gateways of the sensor nodes, they are divided into two different sets, i.e., restricted set (R_{set}, refer equation (4.7)) and open set (O_{set}, refer equation (4.8)) as follows.

$$R_{set}(\text{Restricted Set}) = \{s_2, s_4, s_8\} \text{ and}$$
$$O_{set}(\text{Open Set}) = \{s_1, s_3, s_5, s_6, s_7, s_9, s_{10}, s_{11}, s_{12}\}$$

Initially, all the sensor nodes from the R_{set} are assigned to their corresponding gateway. For example, according to Table 5.1 and the above R_{set} the sensor node s_4 is assigned to g_1, the nodes s_2 and s_8 are assigned to g_2 and no sensor node is assigned to g_3. After assigning all such restricted nodes, the min-heap is built as shown in Figure 5.1.

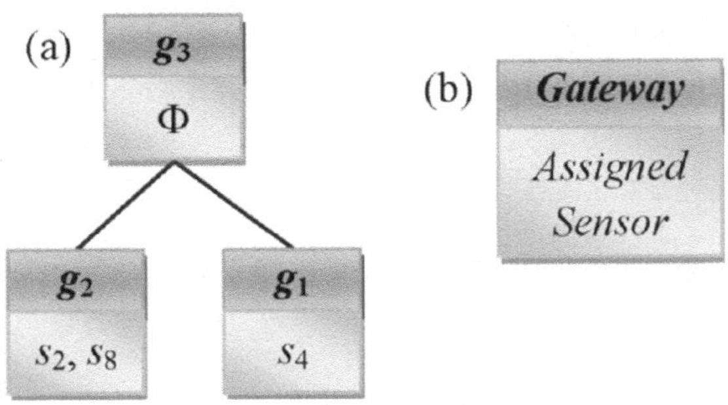

FIGURE 5.1: (a) Min-heap with restricted nodes and (b) min-heap node format

The minimum loaded gateway is found at the root. Therefore, we assign to this gateway that sensor node (from O_{set}), which is nearest to it and within its communication range. It can be seen from Table 5.1 that s_{12} is within communication range and also nearest to g_3. Therefore, s_{12} is assigned to g_3. After assigning the sensor node, the min-heap is rearranged so that the next minimum loaded gateway, i.e., g_1 is at the root which is shown in Figure 5.2(a). Now another sensor node is taken from O_{set} which is nearest to g_1. As Table 5.1 shows that s_3 is nearest to g_1, so s_3 is assigned to g_1. Figure 5.2(b) shows the rearrangement of the min-heap after assigning s_3 to g_1. In the same way, all the sensor nodes can be assigned and the final min-heap can be seen in Figure 5.2(c).

Note: In each step, our main target is to assign a sensor node to the minimum loaded gateway so that the overall load of the gateways can be minimized. Therefore, in order to find out the minimum loaded gateway we need to execute a searching algorithm which is a time consuming process. To solve this problem, we exploit min-heap which helps us to find out the minimum loaded gateway in $O(\log m)$ time (i.e., $O(\log m)$ time to maintain the heap + $O(1)$ time to access the minimum element) rather than $O(m)$ time in searching.

Lemma 5.1. *The time complexity of the algorithm EELBCA is $O(n \log m)$.*

Proof. Step 1 requires $O(n)$ time for assigning Rset to their corresponding gateways. Step 2 requires $O(m \log m)$ time to build a min-heap using m number of gateways. In worst case, step 3 iterates n times in which Step 3.1 through Step 3.3 require constant time and Step 3.4 requires $O(\log m)$ time alone. Therefore, Step 3 can be executed in $O(n \log m)$ time. Thus the above algorithm requires overall time of $O(n \log m)$ as $n \gg m$. □

Lemma 5.2. *The above algorithm EELBCA correctly addresses the issues of load balancing of the CHs and the energy consumption of the sensor nodes.*

Proof. The correctness of the above algorithm can be readily seen from the above

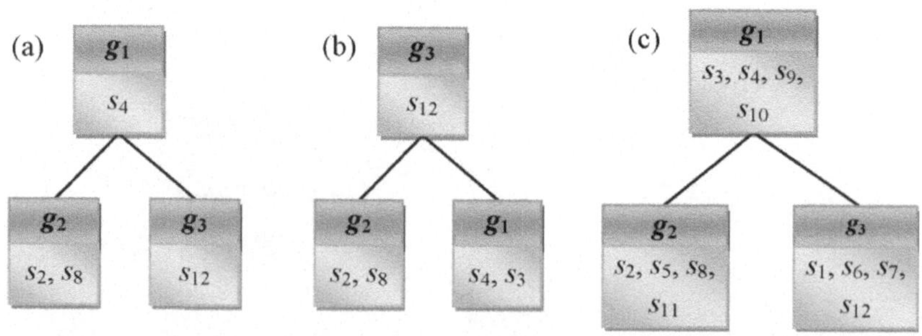

FIGURE 5.2: The min-heap (a) after assigning s_1 to g_3 (b) after assigning s_3 to g_1 and (c) after assigning all the sensor nodes to their corresponding gateway

figurative (Figure 5.2) illustration as follows. It can be observed from the above illustration that, at each assignment we pick up the rooted CH from the min-heap, i.e., the CH which has the minimum number of sensor nodes allotted to it and assign a sensor node to that CH. Therefore, the load is distributed over minimum loaded CHs, thereby balancing the load of the CHs. Energy consumption heavily depends on the distance between two nodes. At each assignment of the sensor nodes, we select and assign that sensor node which is nearest to the minimum loaded CH. Thereby reducing the overall energy consumption of the sensor nodes. Thus the algorithm considers both the issues, i.e., load balancing of the CHs and the energy efficiency of the sensor nodes. □

Theorem 5.1. *The above algorithm EELBCA runs in O(n log m) time by considering the issues of load balancing of the CHs and the energy consumption of the sensor nodes.*

Proof. From Lemma 5.2 and Lemma 5.1. □

5.3 Energy efficient algorithm

As discussed in section 5.2, the algorithm EELBCA performs clustering by energy saving of the sensor nodes and also balancing the load of the gateways in terms of the number of allotted sensor nodes. However, it leaves the communication load of the CHs for transmitting the processed data to the base station. The present algorithm considers all such issues, i.e., gateway load (in terms of processing and communication with sensors and the base station) as well as the energy consumption of the sensor nodes as follows.

The gateways consume their energy to receive the sensed data by its member sensor nodes, aggregate the collected data and finally to send the aggregated data to the base station. If a gateway g_i has n_i number of members, then the total energy consumption of g_i in a single round is as follows

$$E_{Gateway}(g_i) = n_i \times l \times E_{elec} + n_i \times l \times E_{DA} + l \times E_{Transmit}(g_i, BS) \qquad (5.1)$$

where, message size is l-bit. On the other hand, if a sensor node s_p is assigned to gateway g_i, then s_p will consume its energy to transmit the data to g_i. The required transmitting energy, $E_{Transmit}(s_p, g_i)$ can be calculated by the equation (1.1).

Here, $E_{Gateway}(g_i)$ (see equation (5.1)) considers only energy consumption of the gateways and $E_{Transmit}(s_p, g_i)$ considers only energy consumption of the sensor nodes. At the time of cluster formation, by considering only $E_{Gateway}(g_i)$, load of the gateways in terms of energy consumption will be balanced but few sensor nodes may consume more energy due to long haul communication. On the other hand, considering only $E_{Transmit}(s_p, g_i)$, energy consumption of the sensor nodes will be minimized

Algorithm 5 Parameter-Based-Clustering-Algorithm (PBCA)

Input: (1)Set of sensors $S = \{s_1, s_2, ..., s_n\}$.
(2) Set of gateways $\xi = \{g_1, g_2, ..., g_m\}$; where, $m < n$.
(3) $dis(s_i, g_j)$: All distance between s_i to g_j; $g_j \in G_i$.
(4) R_{set}, O_{set} and two parameters, k_1 and k_2.
Output: An assignment $A : S \rightarrow \xi$ such that energy consumption of CHs is balanced and also energy consumption of sensor node is minimized.

while ($R_{set} \neq NULL$) **do**
    Assign successive s_i to their corresponding g_j such that $s_i \in R_{set}$ and delete s_i from R_{set} and S.
end while
Sort the O_{set} in non-decreasing order on the number of possible gateways upon which a sensor node can be assigned.
while ($S \neq NULL$) **do**
    Pick up the first sensor node from O_{set}, say s_i
    Assign s_i to g_k such that such that $Cost(g_k, s_i) = Min\{Cost(g_j, s_i) | \forall g_j \in G_i\}$.
    Delete s_i from O_{set} and S
end while
Stop.

but load of the gateways will not be balanced. Therefore, overloaded gateways consume more energy and die quickly. As a result, the network will be partitioned. So, we have to make a bridge between load balancing and energy consumption of the sensor nodes. This problem can be overcome as follows.

Before assigning a sensor node s_j to a gateway g_k, there may be few sensor nodes which are already assigned to g_k. Let n_k be the number of sensor nodes which have already been assigned to g_k. If s_j is assigned to g_k then the number member sensor nodes of g_k will be $n_k + 1$ and the energy consumption of g_k will be

$$E_{Gateway}(g_i) = (n_k + 1) \times l \times (E_{DA} + E_{elec}) + l \times E_{Transmit}(g_i, BS) \qquad (5.2)$$

On the other hand s_j requires the transmission energy to transmit the sensed data to g_k. The transmitted energy can be calculated from equation (5.2). Before assigning s_j, we calculate the cost of all possible gateways for s_j. We select the gateway with minimum cost value and assign s_j to the gateway. The cost value is calculated in terms of possible consumed energy of all possible gateways of s_j and the sensor node s_j itself. The cost value, $Cost(g_k, s_j)$ for g_k and s_j can be calculated by the cost function as follows.

$$Cost(g_k, s_j) = k_1 \times E_{Gateway}(g_k) + k_2 \times E_{Transmit}(s_j, g_k) \qquad (5.3)$$

Here, two parameters k_1 and k_2 are imposed. By controlling the value of these parameters, load balancing as well as energy consumption of the sensor nodes are controlled. Load is balanced if the parameter k_1 has more value than the second parameter k_2 and energy consumption of sensor nodes is minimized if the parameter k_2 takes higher value than k_1.

Network setup of this algorithm is same as our previous algorithm EELBCA. We first assign the restricted nodes to their corresponding gateway. Now we sort the open set in non-decreasing order on the number of possible gateways upon which the sensor node can be assigned. Let $O_{set} = \{s_a, s_b, s_c \ldots, s_p\}$ be this sorted open set. We now successively consider each sensor node from this sorted list (starting with s_a) to assign it to their correct gateway. In order to assign s_a, we consult its corresponding set of gateways which are within communication range of s_a, i.e., gateways from G_a. Then calculate the cost values using equation (5.3) for all the gateways belongs to G_a and s_a is assigned to that gateway, which has the minimum cost value. The same procedure is continued until all the sensor nodes from O_{set} are allotted to their correct gateway.

Here, the main motivation is to assign that sensor node first which has least chance of getting any gateway, i.e., lesser number of possible gateways within its communication range. Thus at the last of the assignment, the remaining sensor nodes will have comparably higher number of gateways to be assigned. Therefore, there would be more scope to balance the gateway load. The sorting of the open set in non-decreasing order on the number of possible gateways helps to find out the proper order of the sensor assignment. The algorithm is formally presented in Algorithm 5.

Lemma 5.3. *The time complexity of the algorithm PBCA is $O(n \log n)$.*

Proof. Step 1 requires $O(n)$ time for assigning R_{set} to their corresponding gateways. Step 2 requires $O(n \log n)$ time to sort the O_{Set}. Step 3 iterates n times in which Step 3.1 and Step 3.3 each requires constant time and Step 3.2 requires $O(m)$ time alone. Therefore, Step 3 is executed in $O(nm)$ time. Thus the above algorithm requires overall $O(n \log n)$ time assuming $m < \log n$. □

It is observed that the EELBCA performs better than LBCA and LBC in terms of load balancing and execution time. EELBCA uses a single min-heap of m number of nodes to assign n sensor nodes. After each assignment of sensor node, it rearranges the min-heap which takes only $O(\log m)$ time and the overall time complexity of the algorithm is $O(n \log m)$ in worst case. Thus EELBCA consumes very less execution time than LBCA and LBC.

The algorithms are also tested interms of the energy (J) consumption of the network per round. As, both the algorithms LBCA and LBC do not consider distances from sensor nodes to gateways thereby consuming more energy than EELBCA. Performance of the network heavily depends on the number of active gateways, because only gateways can communicate with the base station. As it is already mentioned that PBCA balances the energy consumption of the gateways; it can be observed from Figure 5.3(a) that the rate the gateways die for PBCA is quite less than that of the EELBCA, LBCA and LBC. Figure 5.3(b) shows the energy consumption of the networks against rounds. It can be observed that the energy consumption of PBCA is quite less than LBCA, LBA and EELBCA. The algorithms are also compared in terms of total number of data packets received by the base station. The number of packets delivery to the BS heavily depends on the active gateways in the network. It is observed that the proposed algorithm out performs with respect to number of packet transfer to the BS

We also compare the performance of PBCA by changing the parameter values. We have already mentioned in section 5.3 that between two parameters, first parameter, i.e., k_1 controls the load balancing of the gateways in terms of the energy consumption and the second parameter, i.e., k_2 controls the energy consumption of the sensor nodes. In our simulation, we have taken three cases, i.e., Case I: $k_1 = k_2$, Case II: $k_1 \gg k_2$ and Case III: $k_1 \ll k_2$. Figure 5.4(a)-(b) shows the performances of PBCA under these three cases in terms of the rate at which gateways die and the number of active sensor nodes in the WSN respectively. It can be observed from Figure 5.4(a) that Case II (i.e., emphasizing on load balancing on gateways) performs better than Case I and Case II in terms of the number of active gateways in the network. However, Figure 5.4(b) shows that Case II performs poor in the case of live sensor nodes. This is because; Case II only concentrates on the gateway's life not on the life of the sensor nodes. On the other hand, Case III performs totally reverse of Case II. Case III performs well in sensor node's life but it shows poor performance in gateway's life. It can be noted from Figure 5.4(a)-(b) that Case I always performs in between Case II and Case III, i.e., Case I performs better than Case III in gateway's life and better than Case II in sensor node's life.

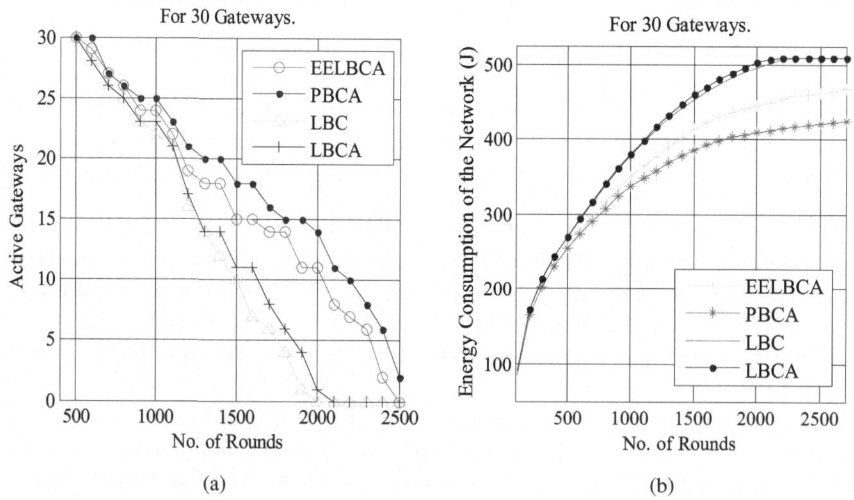

FIGURE 5.3: Comparison in terms of (a) active gateways and (b) consumed energy (J) of the network

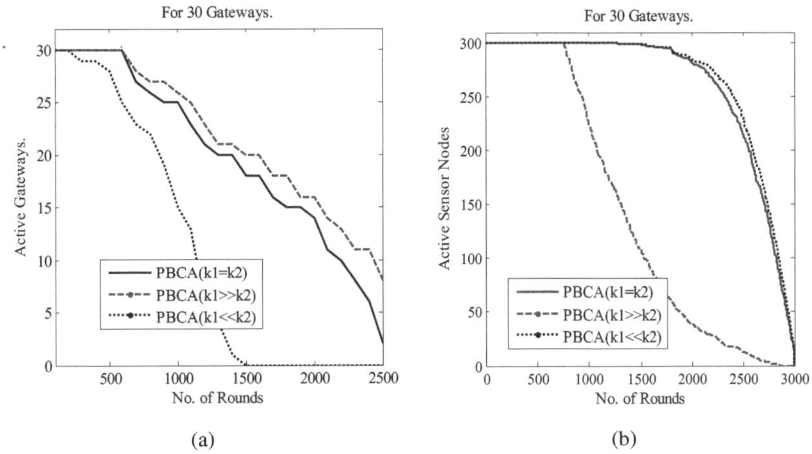

FIGURE 5.4: Comparison of PBCA, by changing the parameter value. (a) active gateways and (b) active Sensor Nodes

---

5.4 Chapter Summary

In this chapter, two clustering schemes have been presented for wireless sensor networks. Both of the algorithms use some high energy gateways as cluster heads (CHs). The algorithms take care of the load balancing of CHs as well as energy efficiency of the sensor nodes. The first algorithm EELBCA has been shown to run in $O(n \log m)$ time in worst case for n sensor nodes and m CHs and the second algorithm PBCA runs in $O(n \log n)$ time in worst case. EELBCA balances the load of the gateways in terms of the number of allotted sensor nodes without considering the communication load of the gateways, whereas PBCA balances the load of the gateways in terms of the overall communication load of the CHs. PBCA is shown to be more energy efficient than the previous one and also prolong network life by preventing the initial death of the unbalanced CHs.

Note that the algorithms perform better in the scenarios where nodes are static after their deployment. Dynamic scenario of the nodes or node mobility is not considered. Moreover, as it is a centralized algorithm, the BS needs the entire network information, which may require high message transfer to the BS at the starting of the network, i.e., in bootstrapping. Therefore, more research can be carried out in the direction of designing a distributed load balanced clustering algorithm with consideration of node mobility.

Chapter 6

Metaheuristic Approaches

Chapter 6

Metaheuristic Approaches

6.1 Introduction

Meta-heuristic algorithms are one of the most popular substitutes for solving NP-hard optimization problems over classical optimization techniques that are based on mathematical and dynamic programming. In fact, due to complex nature of the optimization problems under uncertainty and the high computational complexity, classical approaches that ensure optimal solution are applicable only for small size instance of the problems. In contrast, approaches based on metaheuristic are capable of finding near optimal solutions to problem instances of realistic size, in a generally smaller computation time. Metaheuristics have the following characteristics: 1) they are nature-inspired which are based on some principles from physics, biology or ethology, 2) they make use of stochastic components that involve random variables and 3) they have several parameters that need to be fitted to the problem at hand [142]. Metahueristic algorithms have been successfully applied to solve many NP-hard problems including traveling salesman problem, maximum clique problem, flow shop scheduling problem, and p-median problem. They are also applied in numerous problems associated with feature selection in pattern recognition, automatic clustering, machine learning, and computational biology. Metaheuristic algorithms can be classified as local search vs. global search. Simulated annealing, Tabu search, iterated local search and variable neighborhood search are some popular local search methods. Global search metaheuristics are usually population-based metaheuristics which include evolutionary computation, genetic algorithms, and particle swarm optimization. These algorithms have the following simple generic steps:

1. Initialize a random population of individuals.
2. Apply the search (e.g. mutation and crossover) operators.
3. Select the best individuals to the next generation.
4. Repeat steps 2-3 until a good enough individual is present in the population.

It is mentioned that finding the optimal cluster is an NP-hard problem. Moreover, if m gateways have an average of d valid one-hop neighbor relay nodes, then the number of valid routes is d^m. This implies that the computational complexity of finding the optimal route for a large WSN is very high by a brute force approach. Furthermore, an optimization method requires reasonable amount of memory and computational resources and yet finding out good results is desirable. Therefore, nature-inspired algorithms have drawn enormous attention to solve many optimization problems including clustering and routing in WSNs [57, 58, 78, 79, 143, 144, 145, 146].

The clustering and routing algorithms discussed in this chapter are based on the population-based meta-heuristics only. In this chapter, clustering and routing algorithms are presented based on evolutionary approaches, namely Genetic Algorithm (GA), Differential Evolution (DE) and Particle Swarm Optimization (PSO). The GA based clustering algorithm is developed to solve the load balancing problem as stated in Chapter 4. It differs from the traditional GA [147] with the following respects.

- In the phase of initial population generation, the generation of initial population is restricted by considering the connectivity between the sensor nodes and their CHs. This is in contrast to fully randomized generation of chromosomes as used in the traditional GA.
- In the mutation phase, the mutation point is selected in such a way that it generates children chromosomes that ensures better load balancing. This is in contrast to traditional GA in which mutation point is selected randomly.
- The above two strategies of initial population generation and mutation make the algorithm faster than the traditional GA.

The GA based approach can work for both the equal and unequal load of the sensor nodes. Extensive simulations of the algorithm is performed. The comparison clearly demonstrates that the algorithm performs better than all such algorithms.

Next, the differential evolution (DE) based Clustering Algorithm called DECA is developed with the main objective of prolonging the network life time of the WSNs by taking care of the energy consumption of the common sensor nodes and the gateways (i.e., CHs). Thus it provides the better result than the GA based clustering algorithm. By the network life time, here also it is meant the time from the deployment of the WSN till the death of the first CH. The first death of the CH is delayed through balancing the energy consumption of the CHs which is implemented by the rate of energy consumption and residual energy. In order to find out fast and efficient solution, a local improvement phase is introduced in the traditional DE. This local improvement phase helps the DE based approach to converge faster than the traditional DE and the genetic algorithms (GAs).

Finally, a PSO based clustering and routing algorithms are presented with an efficient particle encoding scheme for complete clustering and routing solution. The PSO-based clustering takes care of energy consumption of the normal sensor nodes as well as the gateways.

6.2 System model and terminologies

Here, the same energy and network models and also the notations are used as described in Section 4.2. However, a few extra notations are used described as follows.

1. $E_{residual}(g_i)$ denotes the remaining energy of g_i.

2. $E_{Gateway}(g_i)$ denotes the total energy consumption of g_i in a single round. The gateways consume their energy to receive the sensed data by their member sensor nodes, data aggregation and finally to send the aggregated data to the base station. If g_i has n_i number of member sensor nodes, then the total energy consumption of g_i in a single round can be calculated as follows

$$E_{Gateway}(g_i) = n_i \times E_R + n_i \times E_{DA} + E_T(g_i, g_{m+1}) \tag{6.1}$$

where E_R, E_{DA} and E_T are the energy consumption due to receiving, data aggregation and transmission to BS respectively.

3. Let $L(i)$ denote the lifetime of the gateway g_i. If g_i has residual energy $E_{residual}(g_i)$ and consumes energy $E_{Gateway}(g_i)$ per round then $L(i)$ can be calculated as follows

$$L(i) = \left\lfloor \frac{E_{residual}(g_i)}{E_{Gateway}(g_i)} \right\rfloor \tag{6.2}$$

4. $AvegClusDist(g_j)$ denotes the average distances from all the member sensor nodes to g_j. Therefore, this is calculated after a complete assignment as follows

$$AvegClusDist(g_j) = \frac{1}{n_j} \sum_{i=1}^{n} \{dis(s_i, g_j) \times \alpha_{i,j}\} \tag{6.3}$$

where n_j denotes the number of member sensor nodes of g_i and $\alpha_{i,j}$ is a Boolean variable such that

$$\alpha_{ij} = \begin{cases} 1 & \text{If sensor node } s_i \text{ assign to cluster head } g_j, \\ & \forall i, j : 1 \leq i \leq n, 1 \leq j \leq m \\ 0 & \text{Otherwise.} \end{cases} \tag{6.4}$$

5. $PNextHops(g_i)$: The set of gateways those might be selected as a next hop relay of g_i. The next hop relay node must be towards the BS. Therefore,

$$PNextHops(g_i) = \{g_j | \forall g_j \in \{\xi + g_{m+1}\} \wedge dis(g_j, g_{m+1}) \leq dis(g_i, g_{m+1})\} \tag{6.5}$$

6. $NextHop(g_i)$ denotes the gateway $g_j, g_j \in PNextHops(g_i)$ which is selected as next-hop relay node from g_i towards BS in data routing phase. Here, the next hop node may be the BS when BS is within communication range of g_i.

7. $HopCount(g_i)$ denotes the number of next hops required to reach to the BS from g_i. If g_i directly communicates with BS, then $HopCount(g_i)$ is one. Therefore, $HopCount(g_i)$ can be recursively defined as

$$HopCount(g_i) = \begin{cases} 1, & NextHop(g_i) = g_{m+1}, \text{(i.e., BS)} \\ 1 + HopCount(g_i), & NextHop(g_i) = g_j, \text{(Other than BS).} \end{cases} \quad (6.6)$$

8. A delay is the time elapsed between the departure of a collected data packet from a source (say g_i) and its arrival to the BS [40]. A delay $D(g_i)$ includes the average value of queuing delay (say d_q) per intermediate data disseminator, transmission delay (say d_t), and propagation delay (say d_p). So, $D(g_i)$ is formulated as

$$\begin{aligned} & D(g_i) = (d_q + d_t + d_p) \times HopCount(g_i) \\ i.e., \quad & D(g_i) = K \times HopCount(g_i) \end{aligned} \quad (6.7)$$

where $K = d_q + d_t + d_p$, which is a constant for a particular network [40]. Therefore, minimizing delay is equivalent to minimizing the hop count.

9. Maximum distance ($MaxDist$) between two nodes in the routing path can be defined by

$$MaxDist = \text{Max}\{dis(g_i, NextHop(g_i)) | \forall i, 1 \leq i \leq m, g_i \in \xi\} \quad (6.8)$$

10. Maximum hop count ($MaxHop$) of the gateways can be defined by

$$MaxHop = \text{Max}\{HopCount(g_i) | \forall i, 1 \leq i \leq m, g_i \in \xi\} \quad (6.9)$$

6.3 Genetic algorithms for clustering

6.3.1 An overview of genetic algorithm

Genetic Algorithm (GA) [114, 118, 138, 147, 148] is an adaptive heuristic method that is widely applied in solving many optimization problems [149, 150, 151, 152, 153, 154, 155]. It begins with a set of randomly generated possible solutions known as initial population . An individual solution is represented by a simple string or an array of genes called chromosome (individual) . The length of each chromosome in a population is equal. Each individual is evaluated by a fitness function to judge its performance. The fitness function is chosen in such a way that an individual provides the result close to the optimal solution.

Once the initial population is generated, two randomly selected chromosomes (parents) are used to produce two child chromosomes by a process called crossover

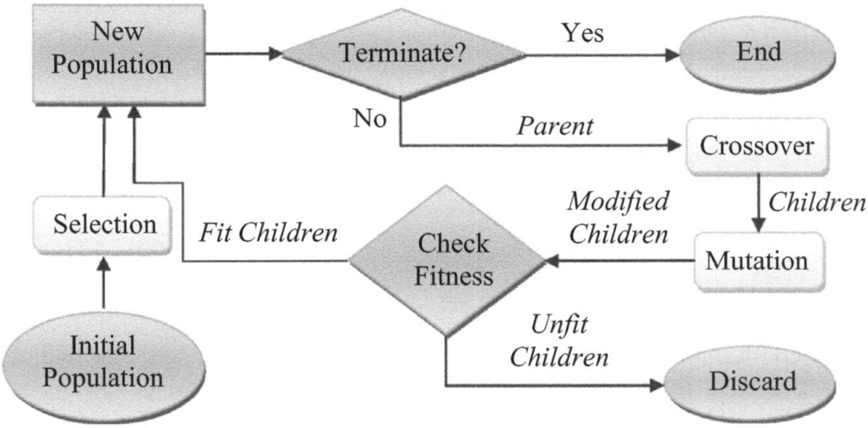

FIGURE 6.1: Flowchart of genetic algorithm

in which the parent chromosomes exchange their genetic information. To produce a better solution, the child chromosomes undergo mutation operation in which their lost genetic values are restored. Whenever the crossover and mutation are over, the fitness function of the child chromosomes is evaluated and their fitness values are compared with that of all the chromosomes of the previous generation. To ensure that the current generation produces better result, two chromosomes of the previous generation with poorest fitness values are replaced with the newly generated child chromosomes. The various steps of a simple GA are depicted in the flowchart as shown in Figure 6.1.

6.3.2 GA based clustering

6.3.2.1 Problem formulation

Let W_j be the load assigned to the cluster head g_j. In other words,

$$W_j = \sum_{i=1}^{n} \{d_i \times \alpha_{ij}\} \tag{6.10}$$

where, α_{ij} is a Boolean variable such that

$$\alpha_{ij} = \begin{cases} 1 & \text{If sensor node } s_i \text{ assign to cluster head } g_j, \\ 0 & \text{Otherwise.} \end{cases} \tag{6.11}$$

Then the overall maximum load of cluster heads is $W = \max \{W_i | \forall g_i \in \xi\}$. Let us address the problem of load balanced clustering, where our main objective is to minimize the overall maximum load of the gateways. The optimization problem of load balanced clustering in terms of Linear Programming (LP) can be formulized as follows.

$$\text{Minimize } W = \max\{W_j | \forall g_j \in \xi\}$$

$$\text{Subject to}$$

$$\sum_{j=1}^{m} \alpha_{ij} = 1 | \forall s_i \in S \tag{6.12}$$

$$\sum_{s_i \in S} d_i \times \alpha_{ij} \leqslant W | \forall g_j \in G_i \tag{6.13}$$

The constraint (6.12) states that a sensor node can be assigned to one and only one gateway and constraint (6.13) indicates that the total load of all the sensor nodes assigned to a gateway must not exceed the overall maximum load of the gateway.

Now, the methodologies for chromosome representation, initial population generation and determination of fitness function are presented followed by selection, crossover and mutation (that are repeatedly invoked) in the subsections as follows [29].

6.3.2.2 Chromosome representation

The chromosome is presented as a string of gateways which indicates the assignment of all the sensor nodes to their corresponding gateway as follows. The length of each chromosome is kept same as the number of sensor nodes. For a chromosome, if i^{th} gene value is say j, then it implies that the sensor node s_i is assigned to the gateway g_j. Note that for any other gene position the same value j can be repeated as more than one sensor nodes can be assigned to the same gateway g_j. The illustration is given with the following example.

Example 6.1. Consider a WSN of 15 sensor nodes and 4 gateways, i.e., $S = \{s_1, s_2, \ldots, s_{15}\}$ and $\xi = \{g_1, g_2, g_3, g_4\}$. So, the length of the chromosome of this network is fifteen. Figure 6.2 shows a chromosome representation, where the gene value at position 7 is 2 and it implies that the sensor node s_7 is assigned to the gateway g_2. Similarly, s_8, s_9 and s_{10} are assigned to g_4, g_3 and g_2 respectively in this representation.

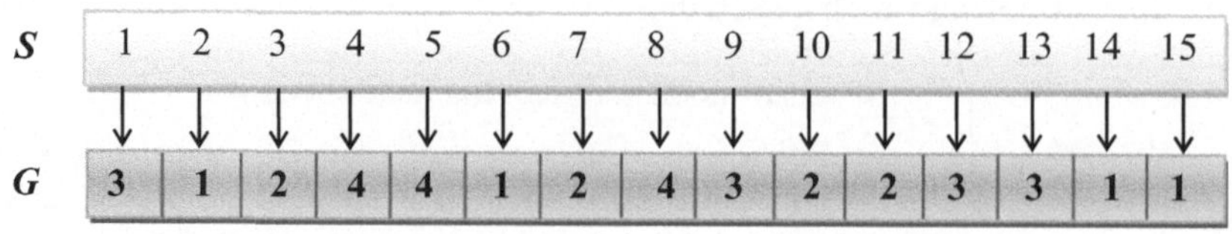

FIGURE 6.2: Chromosome representation

This is important to note that the above chromosome representation is a part of the clustering algorithm. As mentioned above that the length of each chromosome is equal to the number of the sensor nodes, therefore, addition/deletion of any sensor nodes would change the chromosome size and it would require re-clustering.

6.3.2.3 Initial population

The initial population is a randomly generated set of chromosomes. Each chromosome is a sequence of gateways. The valid chromosomes are generated in such a way that the value say j of the ith position gene is randomly selected such that $g_j \in G_i$. It can be noticed that the GA based approach does not depend on any particular algorithm for generating the initial population. It also does not attempt to find a solution which can give a reasonable load balancing. It should also be noted that in the initial population, all of the generated chromosomes represent a complete clustering solution. The idea of generation of an initial population is illustrated with the following example.

Example 6.2. Consider a WSN of 12 sensor nodes and 4 gateways, i.e., $S = \{s_1, s_2, \ldots, s_{15}\}$ and $\xi = \{g_1, g_2, g_3, g_4\}$. Table 6.1 shows all the sensor nodes and their possible gateways to which the sensor nodes can be assigned.

Here, the length of the chromosome is twelve. For the 8th gene position, a number is generated randomly amongst 1, 2 or 4. This is because s_8 can be assigned to any one of the gateways g_1, g_2 or g_4 (refer Table 6.1). In the same way, the 4th gene position can be a randomly generated number amongst 1, 2, 3 or 4 and the 3rd gene position can be any of 2 or 3.

In this example, suppose s_8 selects the gateway g_4 amongst g_1, g_2 and g_4. Similarly, s_4 selects g_3, s_3 selects g_2 and so on. Then with this selection, an individual of the initial population is generated which is shown in Figure 6.3. It should be note

TABLE 6.1: Sensor nodes with the list of possible gateways

Sensor nodes (s_i)	Possible gateways (G_i)
s_1	$G_1 = \{g_1, g_2, g_3, g_4\}$
s_2	$G_2 = \{g_1, g_2\}$
s_3	$G_3 = \{g_2, g_3\}$
s_4	$G_2 = \{g_1, g_2, g_3, g_4\}$
s_5	$G_5 = \{g_2, g_3, g_4\}$
s_6	$G_6 = \{g_1, g_2, g_3, g_4\}$
s_7	$G_7 = \{g_1, g_2, g_3\}$
s_8	$G_8 = \{g_1, g_2, g_4\}$
s_9	$G_9 = \{g_1, g_3\}$
s_{10}	$G_{10} = \{g_2, g_3, g_4\}$
s_{11}	$G_{11} = \{g_1, g_2, g_3, g_4\}$
s_{12}	$G_{12} = \{g_1, g_2, g_3, g_4\}$

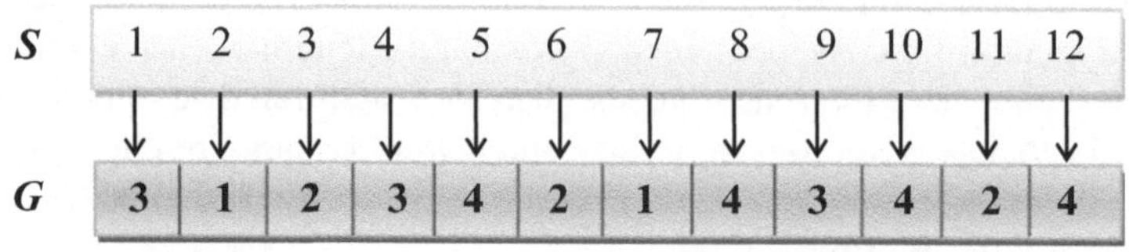

FIGURE 6.3: Generated chromosome from Table 6.1.

that the chromosome shown in this figure represents a complete clustering solution. This is because, the entire twelve sensor nodes are assigned to their corresponding gateway and also a sensor node is assigned to only one gateway.

Remark 6.1. The above strategy of generating initial population makes the algorithm converge faster than the traditional GA. The rationale behind it is that in case of traditional GA, initial population is generated randomly and this may lead generation of many invalid chromosomes which are discarded. This results the process of selection slower.

6.3.2.4 Fitness function

A fitness function is built to evaluate the individual chromosomes of the initial population as follows. Note that the load balancing of the gateway not only minimizes the maximum load of a gateway but also concentrates on the load distribution among all the gateways. Therefore, we construct the fitness function on the basis of the standard deviation (σ) of the gateway load which gives even distribution of the load per cluster. If there are m gateways and n sensor nodes, the standard deviation of gateway load is given by

$$\sigma = \sqrt{\frac{1}{m}\sum_{j=1}^{m}(\mu - W_j)^2}$$

where μ(average load) $= \frac{1}{m}\sum_{i=1}^{n} d_i$, d_i is the load of the sensor node s_i and W_j is the overall load of the gateway g_j. The smaller the standard deviation, the higher is the fitness value. Therefore, the fitness function is chosen as the reciprocal of the standard deviation of the gateway load, i.e. $Fitness = \frac{1}{\sigma}$.

6.3.2.5 Selection

The selection process determines which of the chromosomes from the current population will mate (crossover) to create new chromosomes. For the selection process, we select some valid chromosomes with higher fitness value. The individuals with better fitness values have better chances of selection. There are several selection methods, such as Roulette-wheel selection, rank selection, tournament selection and so on. Tournament selection is used in the method for selecting the chromosomes with best fitness values from the population. The selected chromosomes are

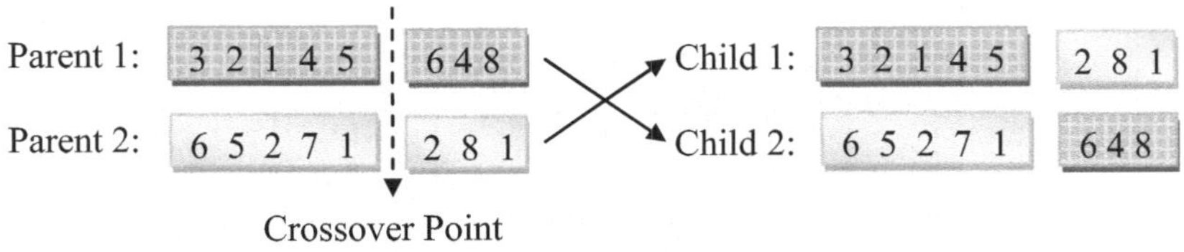

FIGURE 6.4: Crossover Operation

applied to produce new child chromosomes (offspring) by the crossover operation as described in the following section.

6.3.2.6 Crossover

The crossover operation takes place between two randomly selected chromosomes. To produce the new offspring from the selected parent chromosomes, 1-point crossover is used whereby a point is chosen at random, and the two parent chromosomes exchange their information after that point. The whole process is shown in Figure 6.4.

Lemma 6.1. *The two child chromosomes produced by the above crossover operation is valid.*

Proof. A valid chromosome is the one which corresponds to a cluster of the sensor nodes such that for each sensor node the assigned gateway is selected from their gateway list. As mentioned in section 6.3.2.3 , the chromosomes are generated in such a way that the value (say j) of the gene position i is randomly selected so that $g_j \in G_i$. Therefore, for each sensor node the corresponding gateway is valid. □

6.3.2.7 Mutation

The mutation is applied at a selected gene rather than randomly selected gene. Here, our main purpose is to balance the load of the gateways. The particular gateway is selected from the chromosome that has maximum load. As more than one sensor nodes are assigned to the maximum loaded gateway, therefore its number appears in several positions of the chromosome. From all of these gene positions, a gene is randomly selected and it is replaced by another gateway number. It means that we just replace a randomly selected sensor node from that maximum loaded gateway to other less loaded gateway. It can be noted that the replaced number at this gene position must be such that the sensor node is replaced to another gateway from the sensor's gateway list. The idea is illustrated with the figurative example as follows.

Example 6.3. Consider the chromosome representation after crossover operation as shown in Figure 6.5(a) for the same network scenario as example 6.2. It is shown in Figure 6.5(a) that the gateway g_2 is assigned with five sensor nodes and it is maximum. In mutation operation, the method randomly selects a gene position, which is

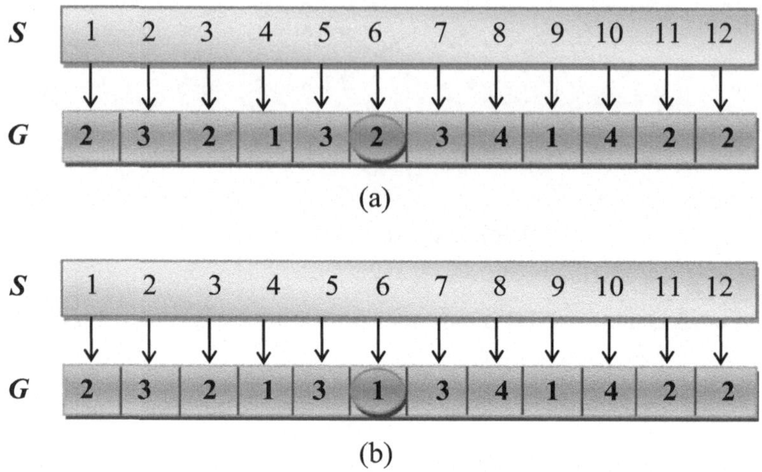

FIGURE 6.5: Chromosome representation (a) Before mutation, (b) After mutation

occupied by the gateway g_2 and replaces it by another gateway number. Note that the replaced gateway must be from the gateway list of the corresponding sensor node. Let a randomly selected gene position be 6 and it is occupied by the gateway number 2. Now, 2 is replaced by another gateway number from G_6 (refer Table 6.1). In this case, a randomly selected gateway number is 1, which is randomly selected amongst 1, 3 and 4 as per the gateway list G_6. The number 2 is not taken as it is the number which has to be replaced. Figure 6.5(b) shows the resultant chromosome after mutation operation.

Lemma 6.2. *The new chromosome produced by the above mutation process is valid.*

Proof. According to the chromosome representation, the maximum loaded gateway lies in more than one gene position of the chromosome. At the time of mutation, the algorithm randomly selects the gene position which is occupied by the maximum loaded gateway number and replaces a new gateway number which is also randomly selected from the gateway list of the corresponding sensor node. Since all valid offspring are generated in crossover operation, the mutation operation cannot hamper the validity of these offspring by replacing a new gateway. □

Remark 6.2. This can be noted that the above strategy of mutation makes the GA based approach to converge faster than the traditional GA. This is because, in case of traditional GA, mutation point is selected randomly. This may lead generation of poor chromosomes and slower convergence rate.

The algorithm is simulated by an initial population of 200 chromosomes. For crossover operation, the best 10% chromosomes using tournament selection are selected. In the simulation, crossover rate was taken as 0.7 and mutation rate as 0.05.

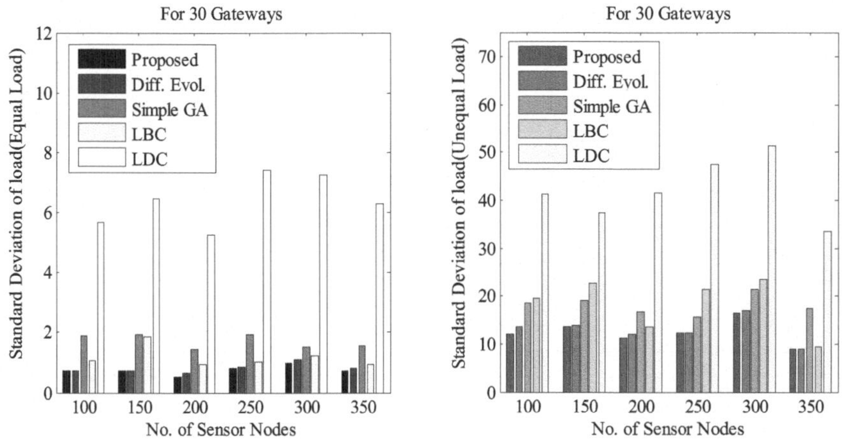

(a) (b)

FIGURE 6.6: Comparison in terms of load balancing for (a) equal and (b) unequal load of the sensor nodes for 30 gateways

The algorithm was run for 75 to 500 iterations. For the comparison purpose, three evolutionary based approaches, i.e., Sajid GA [67], Simple GA [147], Differential Evolution (DE) [156, 157] are run. Two other related clustering algorithms, i.e., LBC [38] and LDC [33] are also run. In the simulation using DE, the donor vector is created by selecting the best chromosome from the population and used the same mutation operation as applied in the presented GA based method. The crossover rate (CR) of the DE was taken 70%. In the simple GA, the fitness function, population size, crossover rate, mutation rate and all other assumptions were taken same as that of the GA.

In order to judge the quality of the load balancing, the standard deviation of the gateway loads is calculated and plotted against the number of sensor nodes. First, the algorithms are run for equal load of the sensor nodes by varying the sensor nodes from 100 to 350 and the number of gateways for 15 and 30. It can be observed that the above GA based method produces better load balancing for equal load of the sensor nodes than others as shown in Figure 6.6(a). Next, the algorithms are run for unequal load of the sensor nodes, the comparison results of which are depicted in Figure 6.6(b).

A sensor node is considered as active if its existing energy is not zero and also there must be at least one gateway within its communication range. Sometimes few CHs die quickly for improper load balancing. As a result, few sensor nodes are unable to find any CH within their range, though the sensor nodes still may have some existing energy. Here, this type of sensor nodes are also considered as inactive. It is observed that the algorithm outperforms simple GA, LDC and the GA based clustering algorithm [67] in terms of number of active sensor nodes. It also performs better than DE and LBC in terms of number of active sensor nodes.

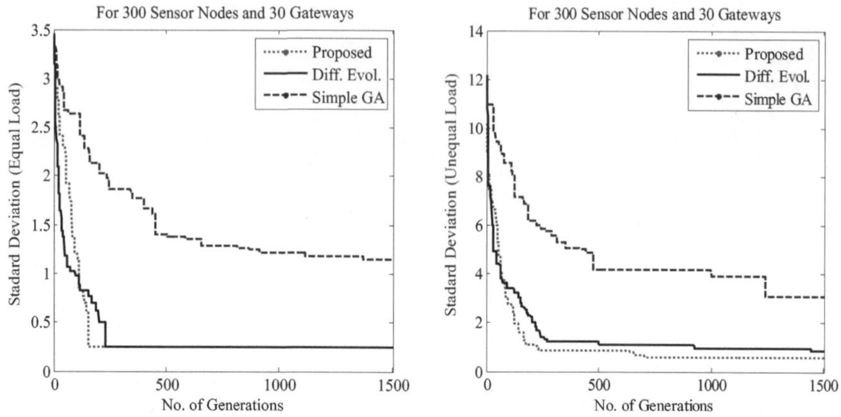

(a) (b)

FIGURE 6.7: Convergence rate of the algorithms for (a) unequal load of the sensor nodes and (b) equal load of the sensor nodes

It should be noted that the algorithm performs better than simple GA, DE, LBC and the GA based clustering in terms of energy consumption. As LDC assign the sensor nodes to their nearest CH, it consumes less energy than other algorithms. However, as it has not considered load balancing issue, all high loaded CHs die quickly and thus many sensor nodes became inactive.

Figure 6.7 shows the comparison of the convergence rate of the GA, simple GA and the DE approach, which clearly shows the faster convergence of the algorithm.

6.4 Differential evolution for clustering and routing

6.4.1 An overview of differential evolution

Differential Evolution (DE) [156, 157] is a stochastic and population based evolutionary algorithm that is widely applied in solving many optimization problems. It consists of four stages, i.e., initialization of population vector, mutation, crossover and selection. DE begins with randomly generated real valued population vectors of some predefined population size (say P_N). The vectors are also known as genome or chromosome and each individual vector gives a complete solution to the multidimensional optimization problem. The dimension D of all the vectors is equal. Each individual vector is evaluated by a fitness function to judge the quality of the solution to the problem.

Once the population vectors are encoded, the algorithm iterates up to T (say) generation with the mutation, crossover and selection operation to enhance the quality

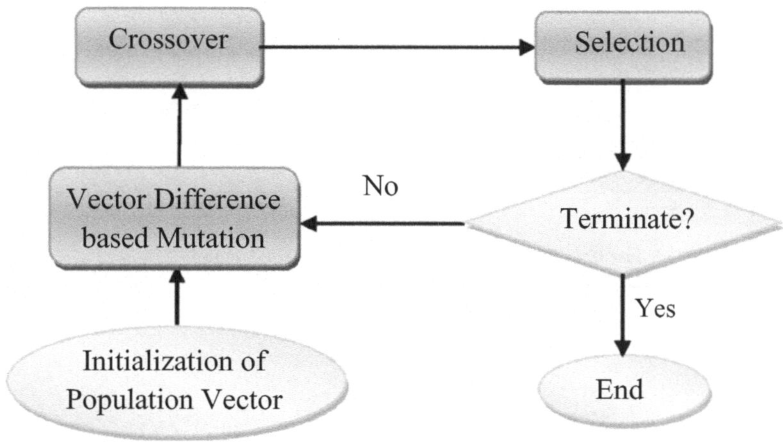

FIGURE 6.8: Flowchart of Differential Evolution

of the population vectors. Therefore, the population vectors are likely to be changed over different generations. Finally, depending on the fitness function, the best vector is selected as the final solution vector. The various steps of a traditional DE are depicted in the flowchart as shown in Figure 6.8. However, in this work, one additional step called local improvement as discussed in section 6.4 is included. The justification of which is given in section 6.4.2.6.

This can be noted that DE has similar computational steps as the traditional evolutionary algorithms (EAs) such as GAs (Genetic algorithms). However, unlike traditional EAs, the DE-variants perturb the current generation population members with the scaled differences of randomly selected and distinct population members. Therefore, no separate probability distribution has to be used for generating the offspring [156].

Various schemes [156, 157] have been proposed for DE. In order to distinguish the schemes, the notation "$DE/x/y/z$" is used, where DE denotes the Differential Evolution, x specifies the vector to be mutated (which can be random or the best vector from the population), y represents the number of difference vectors which are used in the mutation operation and z denotes the crossover scheme (which can be binomial or exponential). Some popular existing DE schemes are $DE/best/1/exp$, $DE/rand/1/bin$, $DE/best/2/exp$, $DE/rand/2/exp$, etc. In the mutation phase of the DE, a donor vector is created for each member vector of the population called target vector . Then the crossover operation takes place between target vector and the donor vector. The donor vector exchanges its components with the target vector to create offspring which is called trial vector. The selection process determines which vector, the target or the trial vector from the current generation (T) survives in the next generation ($T+1$). The selection process keeps the population size constant over subsequence generation. Both of the target and the newly improved trial vectors are

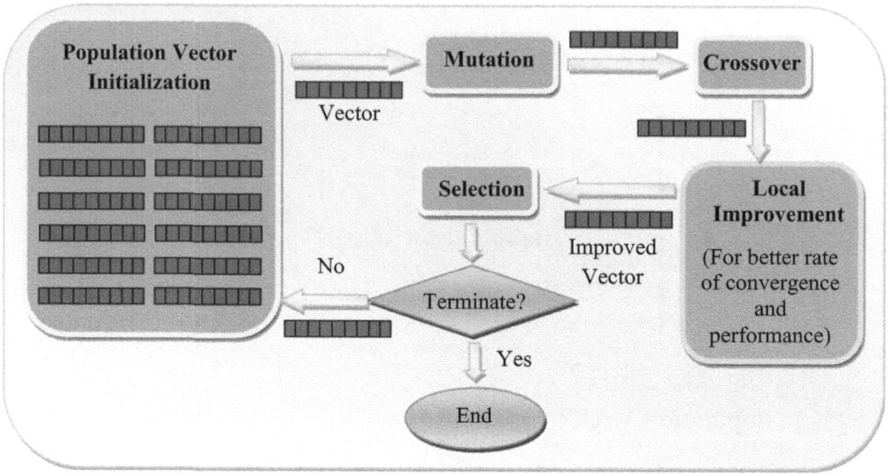

FIGURE 6.9: Flowchart of the DE based clustering algorithm

evaluated by the derived fitness function. The trial vector replaces the target vector if the trial vector has better fitness value; otherwise the target vector retained in the population.

6.4.2 DE based clustering

The above GA based clustering algorithm does not consider residual energy of the gateways which may lead to imbalance energy. Moreover, it does not take care of energy consumption of the sensor nodes. Here, a DE based clustering algorithm called DECA is presented which takes care of te above issues.

6.4.2.1 Problem formulation

For efficient clustering, our objective is to maximize the network lifetime. This can be achieved by maximizing the lifetime of the gateways but reducing the energy consumption of the sensor nodes. In other words, minimize $Z = \frac{A}{L}$, where, $A = Max\{AvegClusDist(g_i), \forall g_i \in \xi\}$ and $L = Min\{L(i), 1 \leq i \leq m\}$.

The flowchart of the DE based clustering algorithm, DECA is depicted in Figure 6.9. This is important to note that the approach has one additional stage of computation, i.e., local improvement. The justification of this extra stage is described in section 6.4.2.6.

Now, the methodologies for population vector initialization and determination of fitness function are presented followed by mutation, crossover, local improvement and selection operation in the subsections as follows [28].

6.4.2.2 Initialization of the population vector

The vectors are presented in such a way that each vector indicates the complete assignment of all the sensor nodes to the gateways. The adopted notation for representing the ith vector of the population at the Tth generation as follows:

$$\overrightarrow{X}_{i,T} = [x_{1,i,T}, x_{2,i,T}, x_{3,i,T}, \cdots, x_{n,i,T}] \tag{6.14}$$

where n is the dimension of the vector and value of the jth component, i.e., $x_{j,i,T}$ maps the assignment of the sensor node s_j to a gateway. Each component is initialized with a randomly generated uniformly distributed number $Rand(0, 1], 0 < Rand(0, 1] \leq 1$. The random number is generated independently for each component. Therefore, the jth component of this vector, i.e., $x_{j,i,T} = Rand(0, 1], 1 \leq j \leq n$ maps a gateway (say g_k) to which the sensor node s_j is assigned. The mapping is done as follows:

$$g_k = Index(G_j, H) \tag{6.15}$$

where $Index(G_j, H)$ is an indexing function that indexes the Hth gateway from G_j and $H = Ceiling(x_{j,i,T} \times |G_j|)$.

This is important to note that the above vector representation is a part of the clustering algorithm. As mentioned above that the dimension of each vector is equal to the number of the sensor nodes, therefore, addition/deletion of any sensor nodes would change the vector dimension and it would require re-clustering.

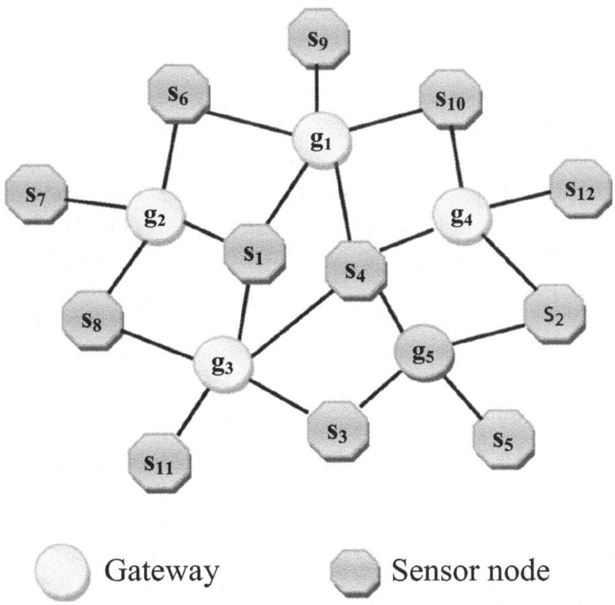

FIGURE 6.10: A Wireless sensor network with gateways

TABLE 6.2: Sensor nodes with the list of possible gateways from figure (6.10)

Sensor Nodes	Possible Gateways (G_j)	$\mid G_j \mid$
s_1	$\{g_3, g_1, g_2\}$	3
s_2	$\{g_4, g_5\}$	2
s_3	$\{g_5, g_3\}$	2
s_4	$\{g_1, g_4, g_5, g_3\}$	4
s_5	$\{g_5\}$	1
s_6	$\{g_1, g_2\}$	2
s_7	$\{g_2\}$	1
s_8	$\{g_3, g_2\}$	2
s_9	$\{g_1\}$	1
s_{10}	$\{g_1, g_4\}$	2
s_{11}	$\{g_3\}$	1
s_{12}	$\{g_4\}$	1

Example 6.4. Consider a WSN with 12 sensor nodes and 5 gateways, i.e., $S = \{s_1, s_2, \cdots, s_{12}\}$ and $\xi = \{g_1, g_2, g_3, g_4, g_5\}$ as shown in Figure 6.10. Therefore, the dimension of the vector is same as the number of sensor nodes, i.e., $N = 12$.

The edges between the sensor nodes and the gateways indicate that the gateways are within communication range of the sensor nodes. It can be observed form Figure 6.10 that the sensor node s_1 is connected with three gateways. In other words, $G_1 = \{g_3, g_1, g_2\}$. The Table 6.2 shows the sensor nodes and the gateways within its communication range.

Now, for each element of the ith vector at the Tth generation, a random number is generated to initialize the element. Let, the generated random number for the first element is 0.46, i.e., $x_{1,i,T} = 0.46$ as shown in Table 6.3. Hence, therefore the 2nd gateway from G_1, i.e., g_1 is selected for assigning s_1 as shown in Table 6.3. In the same way all the sensor nodes are assigned to a gateway using the randomly generated vector. The final assignment of the sensor nodes to their corresponding gateways are shown in Table 6.3.

Therefore, the vector $\overrightarrow{X}_{i,G}$ =[0.46, 0.19, 0.39, 0.67, 0.86, 0.63, 0.24, 0.41, 0.34, 0.73, 0.63, 0.92] maps the assignment of the sensor nodes to their gateways shown in Figure 6.11.

This is important to note that the different order of gateways under G may produce another assignment which is also valid.

6.4.2.3 Derivation of fitness function

Here our main objective is to maximize the network life. This can be possible only if we can make a balance of the life time of the gateways. The general principle behind the balancing of gateway life is as follows:

TABLE 6.3: Sensor nodes assignment from vector encoding

Sensor Nodes	$x_{j,i,T}$	$\mid G_j \mid$	$Ceiling(x_{j,i,T} \times \mid G_j \mid)$	Assigned Gateway
s_1	0.46	3	2	g_1
s_2	0.19	2	1	g_4
s_3	0.39	2	1	g_5
s_4	0.67	4	3	g_5
s_5	0.86	1	1	g_5
s_6	0.63	2	2	g_2
s_7	0.24	1	1	g_2
s_8	0.41	2	1	g_3
s_9	0.34	1	1	g_1
s_{10}	0.73	2	2	g_4
s_{11}	0.63	1	1	g_3
s_{12}	0.92	1	1	g_4

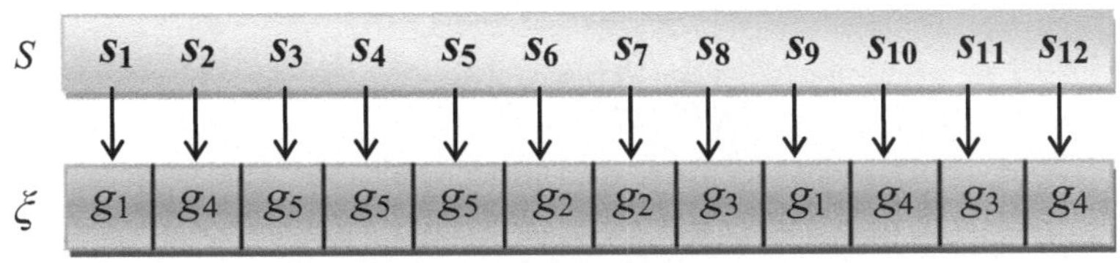

FIGURE 6.11: Final Assignment

Principle: The gateway with lower residual energy should have lower rate of energy consumption per round than the gateways with higher residual energy.

Thus the life time of all the gateways can be balanced effectively. Now, we build the fitness function to evaluate the individual vector depending on the following parameters described as follows.

1. *Standard deviation of lifetime of the CHs*: The CHs consume their energy owing to extra work load from their member sensor nodes and life time of the CHs is very crucial for the long run operation of WSNs. Our main objective is to maximize the minimum life time of the CHs. This can be possible if we can make a balance of the lifetime of the CHs. Balancing the lifetime of the CHs mean that life time of all the CHs is more or less same. This can be possible by applying the above principle. The estimated energy consumption of the gateways and their lifetime can be calculated by the equation (6.1) and (6.2) respectively. In order to judge the quality of the balanced lifetime of the CHs, we measure the standard deviation of the life. The standard deviation (σ_L) can help us to measure the quality of evenly distribution of the gateways life. It can

be calculated as follow.

$$\sigma_L = \sqrt{\frac{1}{m}\sum_{j=1}^{m}\{\mu_L - L(j)\}^2} \tag{6.16}$$

where $\mu_L = \frac{1}{m}\sum_{i=1}^{m} L(i)$ and m is the number of CHs. The shorter the σ_L, the higher is the fitness value. Therefore, the fitness function is reversely proportional to the σ_L. In other words,

$$Fitness \propto \frac{1}{\sigma_L} \tag{6.17}$$

2. *Standard deviation of average cluster distance*: Sensor nodes consume its maximum energy to transmit the sense data to its corresponding CH. In order to minimize the energy consumption of the sensor nodes, they should be assigned to their nearest CH. Here, the objective is to minimize the cluster distance of each cluster. Therefore, we have measured the standard deviation (σ_D) of the cluster distance and it can be calculated as.

$$\sigma_D = \sqrt{\frac{1}{m}\sum_{j=1}^{m}\{\mu_D - AvegClusDist(g_j)\}^2} \tag{6.18}$$

where $\mu_D = \frac{1}{m}\sum_{i=1}^{m} AvegClusDist(g_i)$ and m is the number of CHs. The shorter the σ_D, the higher is the fitness value. Therefore, the fitness function is reversely proportional to the σ_D. In other words,

$$Fitness \propto \frac{1}{\sigma_D} \tag{6.19}$$

Equation (6.17) and (6.19) combinedly implies that

$$Fitness = K \times \frac{1}{\sigma_L} \times \frac{1}{\sigma_D} \tag{6.20}$$

where K is proportionality constant. It is noteworthy that the fitness value is used only comparison purpose. Therefore, the value of K does not hamper the objective. Without loss of generality, we can assume that $K = 1$. Therefore,

$$Fitness = \frac{1}{\sigma_L} \times \frac{1}{\sigma_D} \tag{6.21}$$

$$\text{i.e.,} \quad Fitness = \frac{1}{\sqrt{\frac{1}{m}\sum_{j=1}^{m}\{\mu_L - L(j)\}^2}} \times \frac{1}{\sqrt{\frac{1}{m}\sum_{i=1}^{m} AvegClusDist(g_i)}}$$

$$\text{i.e.,} \quad Fitness = \frac{m}{\sqrt{\frac{1}{m}\sum_{j=1}^{m}\{\mu_L - L(j)\}^2 \times \frac{1}{m}\sum_{i=1}^{m} AvegClusDist(g_i)}}$$

Vectors are evaluated by the above fitness function. Higher is the fitness value, better is the vector.

Lemma 6.3. *The above fitness function takes care of energy consumption of both the gateways and the sensor nodes.*

Proof. The first parameter (σ_L) of the fitness function (equation (6.21)), takes care of the energy consumption of the gateways as follows. The lower the value of σ_L better is the life time of the gateways, (i.e., better is the life time of the network) and this is achieved following the above *principle*. The second parameter σ_D of the fitness function takes care of the energy consumption of the sensor nodes as follows. The lower value of σ_D which minimizes the maximum cluster distance helps to enrich the fitness value. It should be noted that the required transmitting energy of the sensor nodes exponentially proportional to the distances between sensor nodes and the gateways. Thereby, the second parameter reduces the overall energy consumption of the sensor nodes. □

6.4.2.4 Mutation

Here, the scheme *DE/best/1/bin* is used for mutation and crossover operation. For each member vector of the population (called target vector), a donor vector is created through the differential mutation operation by employing the best vector from the population and two other distinct vectors which are generated randomly. Let $\overrightarrow{X}_{i,T}, \overrightarrow{X}_{best,T}$ and $\overrightarrow{V}_{i,T}$ be the target vector, best vector and donor vector respectively. Then the mutation process can be expressed as

$$\overrightarrow{V}_{i,T} = \overrightarrow{X}_{best,T} + F.\overrightarrow{D}_{i,T} \tag{6.22}$$

where $\overrightarrow{D}_{i,T} = (\overrightarrow{X}_{r,T} - \overrightarrow{X}_{s,T})$ and $r, s \in [1, N_P]$ with $i \neq r \neq s \neq best, F$ is the scaling factor which typically lies in the interval [0.4, 1] and is the difference vector of this mutation operation.

It is noteworthy that the above mutation operation is used in standard DE. Unfortunately, it may not work for the scenario. This is because the algebraic subtraction operation of two components of the vectors may cause the resultant components of the difference vector negative. It should be noted that, in the scenario each component of the vectors must be within the range (0,1]. Therefore, the algorithm should generate the components of the difference vector in such a way that it can satisfy the range. This can be made if we choose the component of difference vector as follows

$$d_{j,i,T} = \begin{cases} 1 + (x_{j,r,T} - x_{j,s,T}), & \text{If } (x_{j,r,T} - x_{j,s,T}) \leq 0 \\ (x_{j,r,T} - x_{j,s,T}), & \text{Otherwise} \end{cases} \tag{6.23}$$

As $0 < x_{j,r,T} \leq 1$ and $0 < x_{j,s,T} \leq 1$. The same type of problem may be arisen at the time of arithmetic addition of the mutation process (equation (6.22)). Here, the components of the donor vector are generated as follows

$$v_{j,i,T} = \begin{cases} (x_{j,best,T} + F \times d_{j,s,T}) + 1, & \text{If } (x_{j,best,T} + F \times d_{j,s,T}) > 1 \\ (x_{j,best,T} + F \times d_{j,s,T}), & \text{Otherwise} \end{cases} \tag{6.24}$$

As $0 < x_{j,best,T} \leq 1$ and $0 < x_{j,s,T} \leq 1$.

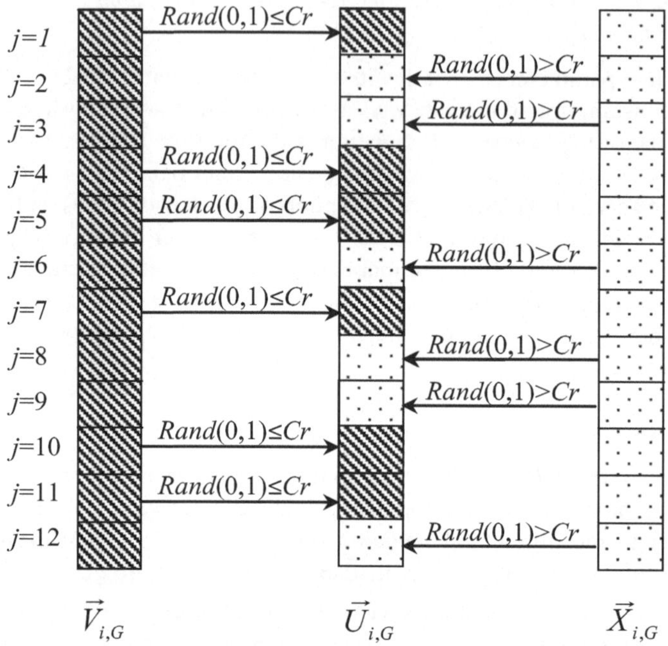

FIGURE 6.12: Crossover operation

6.4.2.5 Crossover

After the mutation operation, crossover operation is performed in between donor vector and target vector to form a new child vector known as trial vector. Here, we use binomial cross over operation with a predefined crossover rate (say Cr). Let $\vec{U}_{i,T} = [u_{1,i,T}, u_{2,i,T}, u_{3,i,T}, \ldots, u_{N,i,T}]$ be the trial vector. Then the crossover operation can be outlined as follows.

$$u_{j,i,T} = \begin{cases} v_{j,i,T} & \text{If } Rand[0,1] \leq Cr \\ x_{j,i,T} & \text{Otherwise} \end{cases} \tag{6.25}$$

To generate each of the components of the trial vector, a number is picked up randomly in between 0 and 1. For the jth component of the trial vector, if the randomly picked up number is less than or equal to the Cr, then j^{th} component of the trial vector is taken same as the jth component donor vector; otherwise, it is taken same as j^{th} component of target vector. The whole crossover process is shown in Figure 6.12.

6.4.2.6 Local improvement

The local improvement step is added to improve the quality of the newly generated trial vector as follows. To perform the local improvement process, we select

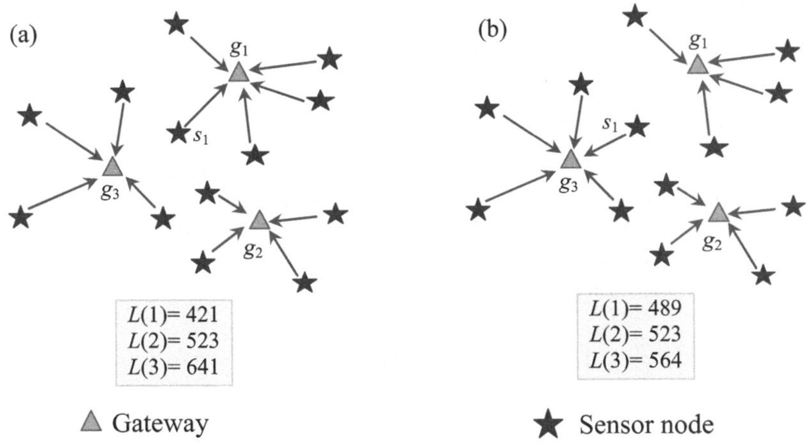

FIGURE 6.13: A WSN of (a) before and (b) after local improvement

the gateway with minimum lifetime from the trial vector. The purpose of selecting this gateway is to enhance its lifetime and thus the network life can be prolonged by delaying the first death of the gateway. This can be possible if we can reduce the load of the gateway by reassigning a member sensor node from its cluster to any other cluster with higher lifetime gateway. To achieve this, a sensor node is randomly selected from the cluster of minimum life gateway and is reassigned to another gateway which is closer to it than the remaining gateways. This is illustrated as follows.

Example 6.5. Consider the WSN as shown in Figure 6.13(a). Assume that this is the configuration of the WSN before the local improvement phase. Also assume that the life time of the gateways g_1, g_2 and g_3 are already calculated and they are $L(1) = 421, L(2) = 523$ and $L(3) = 641$. Therefore, the minimum lifetime gateway is g_1 with 421 rounds. To enhance the lifetime of g_1, a cluster member of g_1 is randomly picked up to reassign another gateway. Let, s_1 be the randomly picked up member sensor node of g_1. Now, s_1 is reassigned to any one of g_2 or g_3 whichever is nearest. Let g_3 be the nearest gateway of s_1. Therefore, s_1 is reassigned to g_3 and the current lifetime of the gateway g_1 is increased but lifetime of g_3 is decreased. Figure 6.13(b) shows this situation of the network after the local improvement phase.

Remark 6.3. The local improvement phase enhances the performance of the algorithm as follows. First the gateway with lowest lifetime is selected and then a sensor node from the cluster of this gateway is randomly picked to reassign the sensor node into another gateway with higher life. Thus lifetime of the lowest life gateway is improved. In other words, first parameter (σ_L) of the fitness function (equation (6.21)) is improved. During sensor reassignment, the randomly selected sensor node is reassigned to the nearest gateway. Therefore, the energy consumption of the reassigned sensor node is also taken care to be minimum. Thus second parameter (σ_D) of the fitness function is also improved. Hence, the local improvement phase helps the trial vector to enhance its fitness and helps the algorithm to converge faster and perform better than the traditional DE and GA.

6.4.2.7 Selection

The selection operation decides which vector from the target vector and newly generated trial vector will survive to the next generation. Both of the vectors are evaluated by the derived fitness function. The trial vector replaces the target vector if the trial vector has better fitness value; otherwise the target vector retained in the population. The selection process is described as

$$\vec{X}_{i,T+1} = \begin{cases} \vec{U}_{i,T} & \text{If } Fitness(\vec{U}_{i,T}) \geq Fitness(\vec{X}_{i,T}) \\ \vec{X}_{i,T} & \text{Otherwise} \end{cases} \tag{6.26}$$

To execute the algorithm DECA, an initial population of 100 chromosomes is considered. In the simulation, crossover rate (Cr) and scaling factor (F) was taken as 0.7 and 0.5 respectively [158]. The algorithm was run for up to 1000 generations.

In order to compare the performance of the DE based algorithm, LBC [38], traditional DE and GA with same fitness function as used the algorithm are also executed. In the traditional DE, the vector encoding schema, fitness function, population size crossover rate and all other parameters were taken same as that of the DE, except the local improvement phase which is introduced in the DE based algorithm. In the experiment using GA, an initial population of 200 chromosomes is considered. In the chromosome generation phase, same encoding schema was used according to the algorithm. For crossover operation, the best 10% chromosomes are selected using

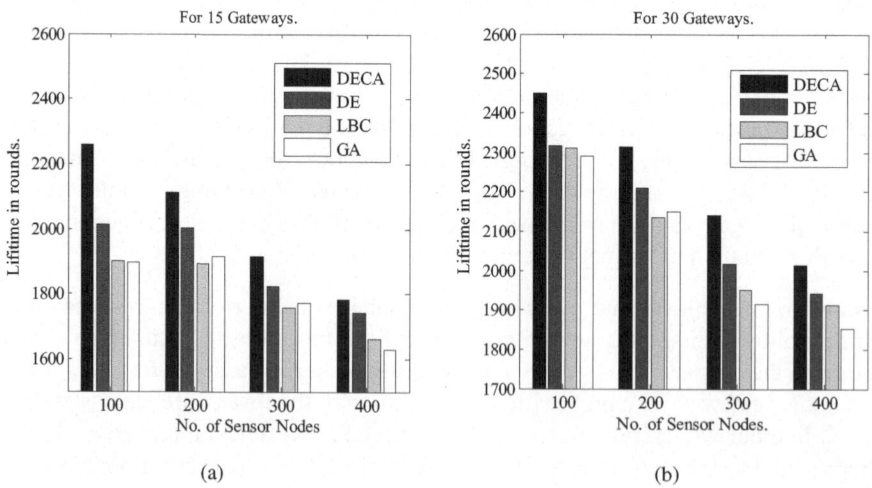

FIGURE 6.14: Comparison in terms of lifetime for (a) 15 and (b) 30 gateways

Roulette-wheel selection and in the simulation, crossover rate was taken as 0.7 and mutation rate as 0.05.

The algorithms are compared in terms of lifetime of the network by varying the sensor nodes from 100 to 400 and the number of gateways for 15 and 30. Figure 6.14 shows the comparison of the DECA, traditional DE, GA and LBC in terms of network life. It can be observed from Figure 6.14 that DECA has better network lifetime than the LBC, traditional DE and the GA based clustering algorithm. Next, the algorithms are compared in terms of the balancing of lifetime of the gateways. Here, the duration between first gateway die (FGD) and last gateway die (LGD) in rounds is calculated and depicted. This is to be noted that lower the duration, better is the balancing of the lifetime. Figure 6.15(a)-(b) shows the comparison. It can be observed from Figure 6.15 that DECA has better balancing than the LBC, traditional DE and GA based clustering algorithm.

Figure 6.16(a)-(b) shows the comparison of the algorithms in terms of number of dead sensor nodes per round and energy (J) consumption of the network respectively. The algorithms are simulated for 300 sensor nodes and 30 gateways. It is clear that, the rate of death of the sensor nodes for DECA, traditional DE and GA is lesser than LBC. This is due to the fact that the derived fitness function takes care about the energy consumption of the normal sensor nodes by reducing the distances between sensor nodes and the gateways. It should be noted that, LBC balances the load of the gateways in terms of number of sensor nodes or cardinality of the cluster. To achieve this goal, some sensor nodes are assigned to a gateway which may be farther from it. As a result their energies are drained out due to long haul transmission and die quickly. The DECA has faster convergence than the traditional DE and the GA, due to the local improvement phase, which is incorporated with the traditional DE.

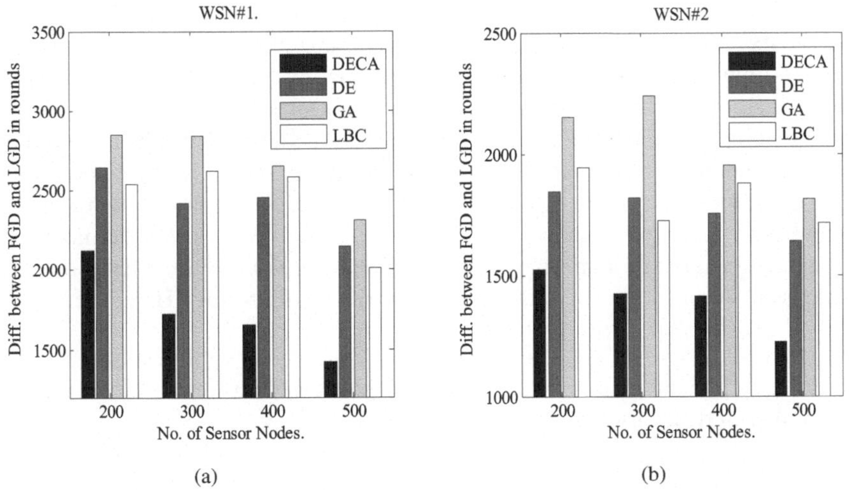

FIGURE 6.15: Difference between FGD (First gateway die) and LGD (Last gateway die) in rounds for (a) WSN#1 and (b) WSN#2

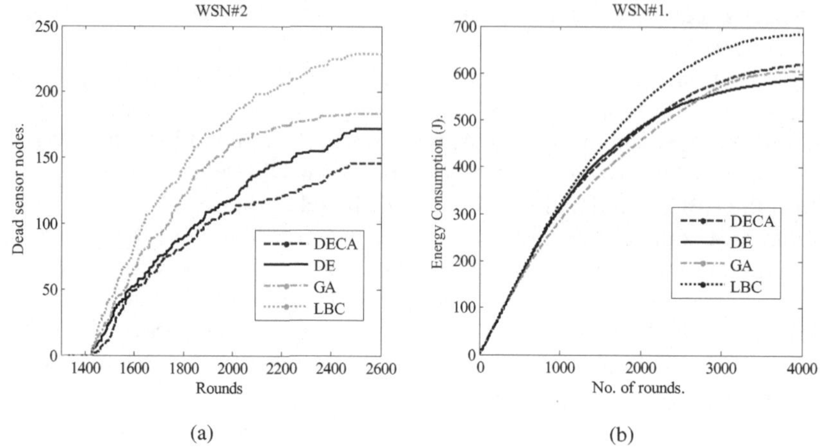

FIGURE 6.16: Comparison in terms of (a) dead sensor nodes and (b) energy consumption of the network

---

6.5 Particle swarm optimization based algorithms

It is noteworthy that the transmission energy (E) which mainly dominates the overall energy consumption of WSNs is proportional to the distance (d) between the transmitter and receiver, i.e., $E \propto d^{\lambda}$, where λ is the path loss exponent and $2 \leq \lambda \leq 4$ [40]. Therefore, minimization of transmission distance can reduce the energy consumption. However, some applications are very time-critical in nature. Hence, they should satisfy strict delay constraints so that the BS can receive the sensed data within a specified time bound. But the delay is proportional to the number of forwards on the dissemination path between a source and the BS [40]. In order to minimize the delay, it is necessary to minimize the number of forwards, which can be achieved by maximizing the distance between consecutive forwards. Therefore, while designing routing algorithms we need to incorporate a trade-off between transmission distance and number of forwards as they pose two conflicting objectives. Furthermore, energy balancing is another important issue for clustering WSN. Particularly, this is a pressing issue when the sensor nodes are not distributed uniformly. Here, even better algorithms are developed to solve the following problems.

1. Energy efficient routing with a trade-off between transmission distance and number of data forwards.

2. Efficient energy balanced clustering for WSNs.

Note that given n sensor nodes and m gateways, the number of possible clusters is m^n. It should also be noted that if the gateways have an average of d valid one-hop neighbor relay nodes, then the number of valid routes is d^m .Therefore the computational complexity of finding the optimal route and cluster for a large WSN seems to be very high by a brute force approach. Moreover, an optimization method requires reasonable amount of memory and computational resources and yet finding out good results is desirable. In order to obtain a faster and efficient solution of the clustering and routing problem with the above issues, a metaheuristic approach such as Particle Swarm Optimization (PSO) is highly desirable. The main objective of this chapter is to develop an efficient PSO-based clustering and routing algorithms for WSNs with the consideration of energy consumption of the sensor nodes for prolonging network life time.

The algorithms are based on another evolutionary approach, i.e., Particle Swarm Optimization (PSO). First, Linear\Non-linear Programming (LP\NLP) formulation is presented for each of the routing and clustering problems separately. Then two PSO-based algorithms for the same are presented. The PSO-based routing builds a trade-off between energy consumption of the CHs and delay in forwarding the data packets. It finds out a route from all the gateways to the base station which has comparably lower overall distance with less number of data forwards. An efficient particle encoding scheme is presented for complete routing solution and design the multi-objective fitness function using weighted sum approach. The PSO-based clustering takes care of energy consumption of the normal sensor nodes as well as the gateways. For clustering, particles are cleverly encoded to produce complete clustering solution. A different fitness function is also used by taking care of those gateways which inevitably consumes more energy by acting as relay node in packet forwarding.

6.5.1 An overview of particle swarm optimization

Particle swarm optimization (PSO) is inspired by natural life, like bird flocking, fish schooling and random search methods of evolutionary algorithm [79, 159]. It can be observed from the nature that animals, especially birds, fishes, etc always travel in a group without colliding. This is because each member follows the group by adjusting its position and velocity using the group information. Thus it reduces individual's effort for searching of food, shelter etc. The various steps of a PSO are depicted in the flowchart as shown in Figure 6.17.

PSO consists of a swarm of a predefined size (say N_P) of particles. Each particle gives a complete solution to the multidimensional optimization problem. The dimension D of all the particles is equal. A particle $P_i, 1 \leq i \leq N_P$ has position $X_{id}, 1 \leq d \leq D$ and velocity V_{id} in the dth dimension of the hyperspace. The adopted notation for representing the ith particle P_i of the population as follows:

$$P_i = [X_{i,1}, X_{i,2}, X_{i,3}, \cdots, X_{i,D}] \tag{6.27}$$

Each particle is evaluated by a fitness function to judge the quality of the solution to the problem. To reach up to the global best position, the particle P_i follows its own best, i.e., personal best called $Pbest_i$ and global best called $Gbest$ to update its own

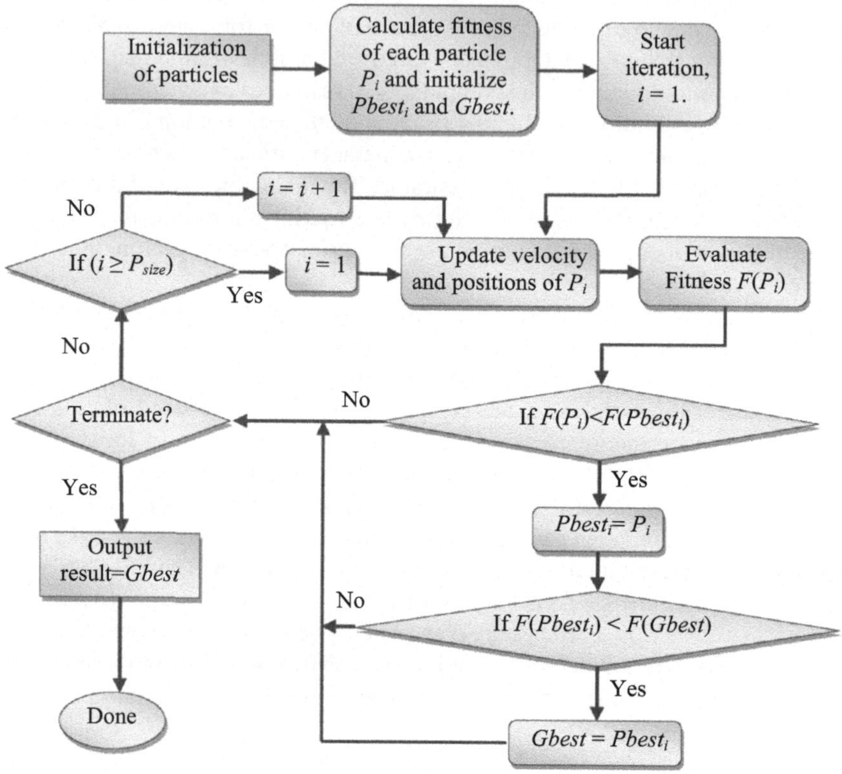

FIGURE 6.17: Flowchart of the PSO (for minimization function)

velocity and position. In each iteration, its velocity V_{id} and position X_{id} in the dth dimension is updated using the following equations respectively.

$$V_{i,d}(t) = w \times V_{i,d}(t-1) + c_1 \times r_1 \times (Xpbest_{i,d} - X_{i,d}(t-1)) + c_2 \times r_2 \times (Xgbest_d - X_{i,d}(t-1)) \quad (6.28)$$

$$X_{i,d}(t) = X_{i,d}(t-1) + V_{i,d}(t-1) \quad (6.29)$$

where w is the inertial weight, c_1 and c_2 are two non negative constants called acceleration factor and r_1 and r_2 are two different uniformly distributed random numbers in the range $[0, 1]$. The update process is iteratively repeated until either an acceptable $Gbest$ is achieved or a fixed number of iterations t_{max} is reached. Particle swarm optimization (PSO) has been applied to solve various optimization problem in manufacturing [143, 146], electrical engineering[160], in the field of WSNs [78, 79], etc.

6.5.2 PSO based routing

6.5.2.1 LP Formulation for Routing Problem

Now, the routing problem is addressed where the main objective is to minimize the maximum transmission distance between two nodes in the routing path and maximum hop count. Let a_{ij} be a Boolean variable defined as

$$a_{ij} = \begin{cases} 1, & \text{If } NextHop(g_i) = g_j \\ 0, & \text{Otherwise} \end{cases} \tag{6.30}$$

Then the Linear Programming (LP) of the routing problem is formulized as follows:

$$\text{Minimize } W = \alpha \times MaxDist + \beta \times MaxHop$$

$$\text{Subject to}$$

$$\sum_{j=1}^{m+1} a_{ij} = 1, 1 \leq i \leq m, \forall g_i \in \xi, \forall g_j \in \{\xi + g_{m+1}\}, i \neq j \tag{6.31}$$

$$dis(g_i, g_j) \times a_{ij} \leq d_{max}, 1 \leq i \leq m, \forall g_i \in \xi, \forall g_j \in \{\xi + g_{m+1}\}, i \neq j \tag{6.32}$$

$$\alpha = 1 - \beta \text{ and } 0 < \beta < 1 \tag{6.33}$$

The constraint (6.31) ensures that the gateway $g_i, \forall i, 1 \leq i \leq m$ forwards its data to only one next hop node g_j and the constraint (6.32) ensures that the selected next hop node is within the transmission range. α and β are two control parameters. α controls the total path distance and β controls the total hop count. The constraint (6.33) defines the range of α and β.

6.5.2.2 Routing algorithm

Now the PSO based routing algorithm is presented. The algorithm consists of particle initialization and determination of fitness function followed by the velocity and position update phase as follows [27].

Initialization of particles:
The particles are represented in such a way that each particle provides the route from each CH to the BS. The dimension of the particles is same and equal to the number of gateways (i.e., m) in the network. Each component, i.e., $X_{i,d}, 1 \leq i \leq N_P, 1 \leq d \leq m$ is initialized with a randomly generated uniformly distributed number $Rand(0, 1], 0 < Rand(0, 1] \leq 1$. The value of the dth component (i.e., $X_{i,d}$) maps a gateway (say g_k) as a next hop relay towards BS from g_d. Therefore, $X_{i,d} = Rand(0, 1]$ maps a gateway (say g_k), indicating that g_d send data to g_k. The mapping is done as follows:

$$g_k = Index(PNextHop(g_d), H) \tag{6.34}$$

where $Index(PNextHops(g_d), H)$ is an indexing function that indexes the Hth gateway from $PNextHops(g_d)$ and $H = Ceiling(X_{i,d} \times |PNextHops(g_d)|)$.

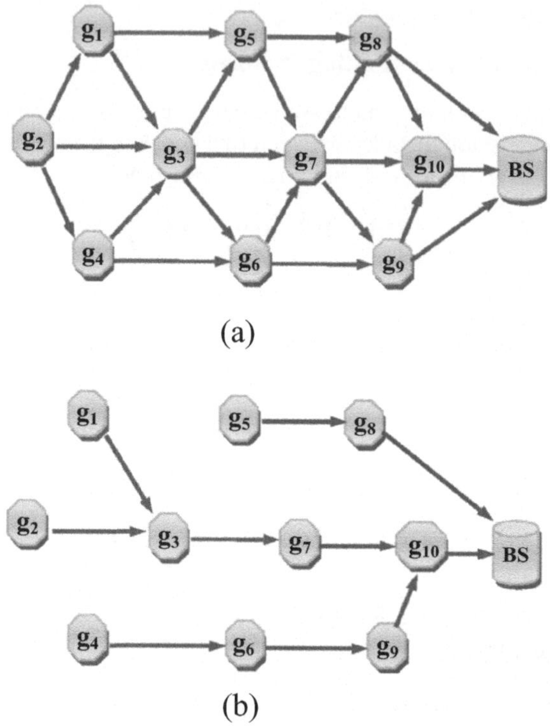

FIGURE 6.18: (a) Sub graph network of a WSN and (b) routing path

Illustration 6.1. Consider a WSN with 10 gateways, i.e., $\xi = \{g_1, g_2, \ldots, g_{10}\}$ as shown in Figure 6.18(a). Therefore, the dimension of the particles is same as the number of gateways, i.e., $m = 10$. Consider the directed acyclic graph $G(V, E)$ shown in Figure 6.18(a), where E represents the set of edges. The edge $g_i \rightarrow g_j$ indicates that g_i can use g_j as a next hop relay towards BS. Here, g_j is closer to BS than g_i and also g_j is within communication range of g_i. It can be observed form Figure 6.18(a) that the gateway g_3 can use any one of the three gateways amongst g_5, g_6, g_7 as a next hop relay node towards the BS. In other words, $PNextHops(g_3) = \{g_5, g_6, g_7\}$. The Table 6.4 shows the gateways and their possible next hop node (gateway/base station) as per the Figure 6.18(a).

Now, for each $X_{i,d}$, $1 \leq d \leq m$ of the particle P_i, a random number is generated to initialize it. Let us assume that a particle P_i = [0.38, 0.63, 0.46, 0.17, 0.86, 0.73, 0.94, 0.81, 0.34, 0.13] has been randomly generated as shown in second column (i.e., $X_{i,d}$) of Table 6.5.

It is shown that this particle actually represents the complete solution of the routing problem as follows. Let consider the second element, i.e., $X_{i,2} = 0.63$. Therefore, which implies that the 2nd gateway from $PNextHops(g_2)$ is selected as next hop relay node of g_2. Thus g_3 (2nd gateway from $PNextHops(g_2)$) serves as the next hop

TABLE 6.4: Gateways and their possible next hops

Gateways	$PNextHops(g_d)$	$\mid PNextHops(g_d) \mid$
g_1	$\{g_3, g_5\}$	2
g_2	$\{g_1, g_3, g_4\}$	3
g_3	$\{g_5, g_7, g_6\}$	3
g_4	$\{g_6, g_3\}$	2
g_5	$\{g_7, g_8\}$	2
g_6	$\{g_7, g_9\}$	2
g_7	$\{g_8, g_9, g_{10}\}$	3
g_8	$\{g_{10}, BS\}$	2
g_9	$\{g_{10}, BS\}$	2
g_{10}	$\{BS\}$	1

TABLE 6.5: Next-hop relay node selection from the randomly generated particle

Gateway	$x_{i,d}$	$\mid PNextHops(g_d) \mid$	$Ceiling(x_{i,d} \times \mid PNextHops(g_d) \mid)$	$PNextHops(g_d)$
g_1	0.38	2	1	g_3
g_2	0.63	3	2	g_3
g_3	0.46	3	2	g_7
g_4	0.17	2	1	g_6
g_5	0.86	2	2	g_8
g_6	0.73	2	2	g_9
g_7	0.94	3	3	g_{10}
g_8	0.81	2	2	BS
g_9	0.34	2	1	g_{10}
g_{10}	0.13	1	1	BS

relay for g_2. In the same way all the gateways are given a next hop relay using the randomly generated particle. Thus, the above randomly generated particle P_i maps the complete routing solution from each CH as shown in fifth column (i.e., *NextHop*) of Table 6.5. Therefore, the route from g_1 to the base station can be expressed as the path $g_1 \rightarrow g_3 \rightarrow g_7 \rightarrow g_{10} \rightarrow BS$. The final routing schedule for the sub-graph network is shown in Figure 6.18(b).

Derivation of fitness function:

Now, a fitness function is constructed to evaluate the individual particle of the population. This helps us to periodically update the personal best and global best of the particles. Here, two objectives are considered. The first objective is to minimize the maximum distance between two nodes and the second objective is to minimize the maximum number of hops used by the gateways. Therefore, the two objectives are as follows.

$$\text{Objective 1: Minimize } MaxDist = \text{Max}\{dis(g_i, NextHop(g_i))|\forall i, 1 \leq i \leq m\} \tag{6.35}$$

$$\text{Objective 2: Minimize } MaxHop = \text{Max}\{HopCount(g_i)|\forall i, 1 \leq i \leq m\} \tag{6.36}$$

It is noteworthy that the above two objectives conflict each other, i.e., lower distance of next hop relay node increases the hop count and vice versa. Therefore, optimization of one objective hampers in optimization of other. The work constructs the fitness function in such a way that a trade-off can be built with these conflicting objectives. Here, weight sum approach (WSA)[161] is used for the construction of the multi objective fitness function. WSA is a classical approach for solving the multi-objective optimization problem. In this approach, a weight value w_i is multiplied with each objective. Finally all the multiplied values are added to convert the multi objectives into a single scalar objective function as follows.

$$Fitness = w_1 \times MaxDist + w_2 \times MaxHops \tag{6.37}$$

In this approach it is taken $w_2 = 1 - w_1$ and $0 < w_1 < 1$. Our objective is to minimize the $Fitness$ value. In other words,

$$\text{Objective: Minimize } Fitness \tag{6.38}$$

Therefore, lower the fitness value, the better is the particle position.

Velocity and position update: The velocity and the position are updated in each iteration using equation (6.28) and (6.29) respectively. It is noteworthy that, the algebraic steps of addition and subtraction operation in equation (6.28) and (6.29) may cause the new position of the particle to be negative or greater than one. In the scenario the position of the particle must satisfy the range (0,1]. Therefore, the algorithm should generate the positions of the particles in such a way that it can satisfy the range. This can be made if we choose the positions as follows.

- If new position is negative or zero, then replace the position value by a newly generated random number which tends to zero.
- If new position is greater than one, then replace the position value by one.

After getting the new position, the particle P_i is evaluated by the fitness function. Now, its personal best ($Pbest_i$) is replaced by itself, only if its current fitness value is better than its $Pbest_i$ fitness value. The updating process is as follows.

$$Pbest_i = \begin{cases} P_i, & if(Fitness(P_i) < Fitness(Pbest_i)) \\ Pbest_i, & \text{Otherwise.} \end{cases} \tag{6.39}$$

Now, the global best is also updated as follows.

$$Gbest = \begin{cases} P_i, & if(Fitness(P_i) < Fitness(Gbest)) \\ Gbest, & \text{Otherwise.} \end{cases} \tag{6.40}$$

Algorithm 6 PSO-Routing

Input: (1) Set of cluster heads $\xi = \{g_1, g_2, ..., g_m\}$.
(2) $PNextHops(g_i)$ and $HopCount(g_i), \forall i, 1 \leq i \leq m$.
(3) Predefined swarm size N_P.
Output: Route : $\xi \rightarrow \{\xi + g_{m+1}\}$.

Initialize particles $P_i, \forall i, 1 \leq i \leq N_P$. /*As described in section 6.5.2.2*/
for (i = 1 to N_P) **do**
    Calculate $Fitness(P_i)$ /* Using equation (6.37) */
    $Pbest_i = P_i$
end for
$Gbest = \{Pbest_k | Fitness(Pbest_k) = min(Fitness(Pbest_i), \forall i, 1 \leq i \leq N_P)\}$
while (!(Terminate)) **do**
    for (i = 1 to N_P) **do**
        Update velocity and position of P_i using equation (6.28) and (6.29)
        Calculate $Fitness(P_i)$
        if ($Fitness(P_i) < Fitness(Pbest_i)$) **then**
            $Pbest_i = P_i$
        end if
        if ($Fitness(Pbest_i) < Fitness(Gbest)$) **then**
            $Gbest = Pbest_i$
        end if
    end for
end while
Stop.

The velocity and the positions are iteratively updated until the termination criteria are fulfilled. In this approach, the termination criterion is a predefined iteration number. After termination of the PSO-based routing algorithm, the particle $Gbest$ represents the final routing solution. The PSO based routing algorithm is shown in Algorithm 6.

6.5.3 PSO Based Clustering

6.5.3.1 NLP Formulation for clustering problem:

Now, the clustering problem is addressed where the basic objective is to maximize the life-time of the network as well as minimize the energy consumption of the sensor nodes. By the network life time, it is meant the time from the deployment of the WSN till the death of the first gateway. Therefore, network life can be maximized if we can maximize the minimum life time of the gateways. Energy consumption of the sensor nodes can be minimized by minimizing the distance between sensor nodes and their corresponding gateways. Let b_{ij} be a Boolean variable such that

$$b_{ij} = \begin{cases} 1, & \text{If sensor node } s_i \text{ is assigned to gateway } g_j \\ & \forall i, j : 1 \leq i \leq n, 1 \leq j \leq m \\ 0, & \text{Otherwise} \end{cases} \tag{6.41}$$

Let L be the minimum lifetime of the gateways, i.e., $L = Min\{L(i)|\forall i, 1 \leq i \leq m\}$ and $AvegDist$ be the average distance between sensor nodes and their corresponding CH, i.e.,

$$AvegDist = \frac{1}{n} \sum_{i=1}^{n} \sum_{j=1}^{m} dis(s_i, g_j) \times b_{ij} \tag{6.42}$$

Then the Non-Linear Programming (NLP) of the clustering problem can be formulized as follows:

$$\text{Maximize } Z = \frac{L}{AvegDist}$$

Subject to

$$\sum_{j=1}^{m} b_{ij} = 1, 1 \leq i \leq n \tag{6.43}$$

$$dis(s_i, g_j) \times b_{ij} \leq R_S, 1 \leq i \leq n, \forall s_i \in S, \forall g_j \in \xi \tag{6.44}$$

The constraint (6.43) states that the sensor node $s_i, \forall i, 1 \leq i \leq n$ can be assigned to one and only one gateway. The constraint (6.44) ensures that the sensor nodes are assigned to the gateway within its communication range.

6.5.3.2 Clustering algorithm

After executing the above routing algorithm, the base station executes the clustering algorithm in which the information of the routing solution is used for the cluster formation to balance the load of the CHs (i.e., gateways). Note that here the used particle initialization for clustering and the fitness function are different from that are used in the above routing algorithm as they do not fit for the clustering.

Initialization of particles:

Here, the dimension of the particle is same as the number of sensor nodes (i.e., n) in the network. Let, $P_i = [Y_{i,1}, Y_{i,2}, Y_{i,3}, \ldots, Y_{i,n}]$ be the ith particle of the population where each component, $Y_{i,d}, 1 \leq i \leq N_P, 1 \leq d \leq n$ maps the assignment of the sensor node sd to a gateway. Each component is initialized with a randomly generated uniformly distributed number $Rand(0, 1], 0 < Rand(0, 1] \leq 1$. The random number is generated independently for each component. The component of the dth dimension of this particle, i.e., $Y_{i,d} = Rand(0, 1], 1 \leq d \leq n$ maps a gateway (say g_k) to which the sensor node s_d is assigned. The mapping is done as follows:

$$g_k = Index(G_d, H) \tag{6.45}$$

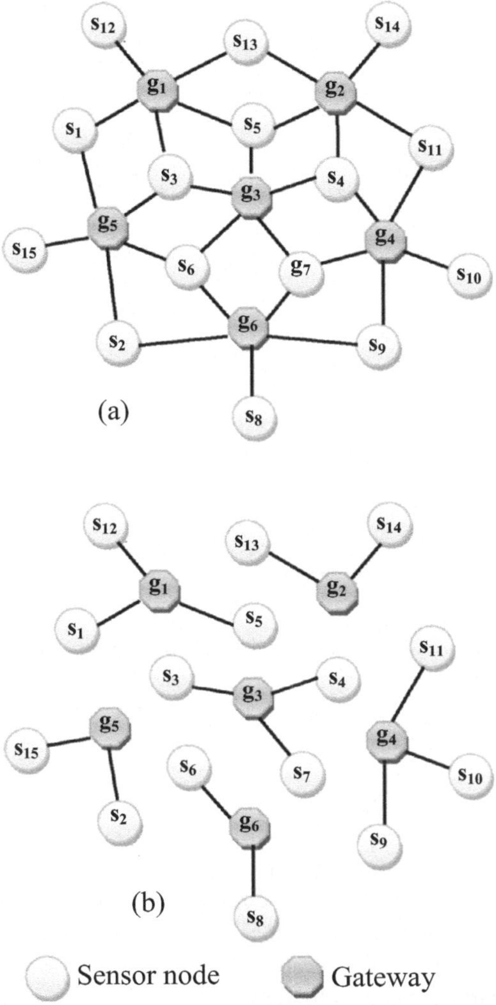

FIGURE 6.19: A WSN with gateways (a) before clustering and (b) after clustering

where $Index(G_d, H)$ is an indexing function that indexes the Hth gateway from G_d and $H = Ceiling(Y_{i,d} \times |G_d|)$.

This is important to note that the above particle representation is a part of the clustering algorithm. As mentioned above that the dimension of each particle is equal to the number of the sensor nodes. Therefore, addition/deletion of any sensor node would change the particle dimension and require re-clustering.

Illustration 6.2. Consider a WSN with 15 sensor nodes and 6 gateways, i.e., $S = \{s_1, s_2, \ldots, s_{15}\}$ and $\xi = \{g_1, g_2, g_3, g_4, g_5, g_6\}$ as shown in Figure 6.19(a). Therefore, the dimension of the particle is same as the number of sensor nodes, i.e., $n = 15$.

TABLE 6.6: Sensor nodes with the list of possible gateways

Sensor nodes	G_d	$\mid G_d \mid$
s_1	$\{g_1, g_5\}$	2
s_2	$\{g_6, g_5\}$	2
s_3	$\{g_1, g_5, g_3\}$	3
s_4	$\{g_3, g_4, g_2\}$	3
s_5	$\{g_3, g_2, g_1\}$	3
s_6	$\{g_6, g_5, g_3\}$	3
s_7	$\{g_4, g_3, g_6\}$	3
s_8	$\{g_6\}$	1
s_9	$\{g_4, g_6\}$	2
s_{10}	$\{g_4\}$	1
s_{11}	$\{g_2, g_4\}$	2
s_{12}	$\{g_1\}$	1
s_{13}	$\{g_1, g_2\}$	2
s_{14}	$\{g_2\}$	1
s_{15}	$\{g_5\}$	1

The edges between the sensor nodes and the gateways indicate that the gateways are within communication range of the sensor nodes. It can be observed from Figure 6.19(a) that the sensor node s_4 is connected with three gateways. In other words, $G_4 = \{g_3, g_4, g_2\}$. The Table 6.6 shows the sensor nodes and the gateways within its communication range.

Now, for each element of the ith particle at each generation, a random number is generated to initialize the element. Let, the generated random number for the fourth element is 0.18, i.e., $Y_{i,4} = 0.18$ as shown in Table 6.7. Hence, $Ceiling(Y_{i,4} \times |G_4|) = 1$ therefore the 1st gateway from G_4, i.e., g_3 is selected for assigning s_4 as shown in Table 6.7. In the same way all the sensor nodes are assigned to a gateway using the randomly generated particle. The final assignment of the sensor nodes to their corresponding gateways are shown in Table 6.7.

Therefore, the particle P_i = [0.26, 0.86, 0.91, 0.18, 0.77, 0.12, 0.47, 0.62, 0.24, 0.53, 0.71, 0.46, 0.81, 0.92, 0.39] maps the assignment of the sensor nodes to their gateways shown in Figure 6.19(b). This is important to note that the different order of gateways under G may produce another assignment which is also valid.

Derivation pf fitness function:

The fitness function is derived in such a way that it takes care of energy consumption of the CHs as well as the sensor nodes. The derivation depends on some parameters described as follows.

1. ***Lifetime of the CHs:*** Our first objective is to maximize the network life. This can be possible if we can maximize the lifetime of the gateway that has least lifetime. The general principle behind the maximization of the gateway life is that the gateway with lower residual energy should have lower rate of energy

TABLE 6.7: Sensor nodes assignment from particle representation

Sensor Nodes (s_d)	$Y_{i,d}$	$\mid G_d \mid$	*Ceiling* ($Y_{i,d} \times \mid G_d \mid$)	Assigned Gateway
s_1	0.26	2	1	g_1
s_2	0.86	2	2	g_5
s_3	0.91	3	3	g_3
s_4	0.18	3	1	g_3
s_5	0.77	3	3	g_1
s_6	0.12	3	1	g_6
s_7	0.47	3	2	g_3
s_8	0.62	1	1	g_6
s_9	0.24	2	1	g_4
s_{10}	0.53	1	1	g_4
s_{11}	0.71	2	2	g_4
s_{12}	0.46	1	1	g_1
s_{13}	0.81	2	2	g_2
s_{14}	0.92	1	1	g_2
s_{15}	0.39	1	1	g_5

consumption per round than the gateways with higher residual energy. Thus the lifetime of the gateways with lesser remaining energy can be prolonged effectively. The gateways consume their energy for receiving sensed data from their member sensor nodes, aggregation of data and finally to send the aggregated data to the base station. Therefore, energy consumption of a gateway g_i with n_i number of member sensor nodes due to inter-cluster activity in a single round can be formulated as follows.

$$E_{CLUSTER}(g_i) = n_i \times E_R + n_i \times E_{DA} + E_T(g_i, NextHop(g_i)) \qquad (6.46)$$

where E_R, E_{DA} and E_T are the energy consumption due to data receiving, aggregation and transmission to BS respectively. Apart from these inter-cluster activities, in a multi-hop scenario, g_i also consumes its energy due to forwarding the data from the gateways whose routing path to the base station goes through g_i. Before calculating the energy consumption due to data forwarding, we need to calculate the total number of incoming packets those are coming from other gateways to g_i toward the base station. It can be recursively calculated as follows.

$$N_{IN}(g_i) = \begin{cases} 0, \text{If } NextHop(g_j) \neq g_i, \forall g_j \in \xi. \\ \sum\{N_{IN}(g_j) | NextHop(g_j) = g_i, \forall g_j \in \xi\}, \text{ Otherwise.} \end{cases} \qquad (6.47)$$

The gateway g_i having $N_{IN}(g_i)$ number of incoming packets will consume its energy for receiving and transmitting these packets. The overall data forwarding energy consumption can be calculated as follows.

$$E_{FORWARD}(g_i) = N_{IN}(g_i) \times E_R + N_{IN}(g_i) \times E_T(g_i, NextHop(g_i)) \quad (6.48)$$

Therefore, total energy consumption of g_i can be calculated by adding equation (6.46) and equation (6.48) as follows.

$$\begin{aligned} E_{Gateway}(g_i) &= E_{CLUSTER}(g_i) + E_{FORWARD}(g_i) \\ &= n_i \times E_R + n_i \times E_{DA} + E_T(g_i, NextHop(g_i)) \\ &\quad + N_{IN}(g_i) \times E_R + N_{IN}(g_i) \times E_T(g_i, NextHop(g_i)) \\ &= (n_i + N_{IN}(g_i)) \times E_R + n_i \times E_{DA} \\ &\quad + (N_{IN}(g_i) + 1) \times E_T(g_i, NextHop(g_i)) \end{aligned} \quad (6.49)$$

Let, g_i has the residual energy of $E_{residual}(g_i)$. Then, lifetime of g_i can be calculated as

$$L(i) = \left\lfloor \frac{E_{residual}(g_i)}{E_{Gateway}(g_i)} \right\rfloor \quad (6.50)$$

Our first objective is to maximize the minimum lifetime of the CHs, i.e.,

$$\text{Objective 1: Maximize } L = Min\{L(i) | \forall i, 1 \leq i \leq m\} \quad (6.51)$$

Therefore, larger the L, higher is the fitness value. In other words,

$$Fitness \propto L \quad (6.52)$$

Remark 1. In the equation (6.47) and (6.48), we have calculated the number of incoming packets those are coming from the other gateways towards base station over the gateway g_i and the corresponding energy consumption respectively. As our first objective (6.51) is to maximize the lifetime of the network, the prior information of the data forwarding overhead of the gateways helps in the clustering phase to properly balance the load of the gateways and prolong the network life. Thus we have takes care of the gateways those inevitably deplete their energy due to serving as relay node in data routing phase.

2. ***Average cluster distance:*** In order to maximize the lifetime of the gateways, some sensor nodes are forced to be assigned to the gateway which is farther from it. Thus these sensor nodes consume their energy faster and die quickly due to long haul communication with their CH. Therefore, we should also take care about the assignment of the sensor nodes to minimize their energy consumption. In order to minimize the energy consumption of the sensor nodes, they should be assigned to their nearest CH. Therefore, we have measured the average distance between sensor nodes and their corresponding CH and the second objective is to minimize it. In other words,

$$\text{Objective 1: Maximize } AvegDist = \frac{1}{N} \sum_{i=1}^{N} dis(s_i, CH_i) \quad (6.53)$$

where, CH_i be the cluster head of sensor node s_i. The shorter the $AvegDist$, the higher is the fitness value. Therefore, the fitness function is reversely proportional to the $AvegDist$, i.e.,

$$Fitness \propto \frac{1}{AvegDist} \tag{6.54}$$

Equation (6.52) and (6.54) combinedly implies that

$$Fitness \propto \frac{L}{AvegDist}$$

$$i.e., \qquad Fitness = K \times \frac{L}{AvegDist} \tag{6.55}$$

where K is proportionality constant. It is noteworthy that the fitness value is used only for comparison purpose. Therefore, the value of K does not hamper the objective. Without loss of generality, it can be assumed that $K = 1$. Therefore,

$$Fitness = \frac{L}{AvegDist} \tag{6.56}$$

Our objective is to maximize the Fitness value. In other words,

$$\text{Objective: Maximize } Fitness \tag{6.57}$$

Therefore, higher the fitness value, the better is the particle position.

Velocity and position update:
In the same way as routing algorithm, the velocity and the position are updated in each iteration using equation (6.28) and (6.29) respectively. Then same way $Pbest_i$ and $Gbest$ are also updated. The clustering algorithm can be developed in the same way as routing algorithm (refer Figure 6).

Note that, in the above PSO based algorithms, after execution of the routing algorithm, the base station uses the final route setup for proper formation of the cluster. When the routing and clustering is over, all the gateways are informed about their next hop relay node towards the base station and the sensor nodes are also informed about the ID of the gateway they belong to. Then the gateways provide a TDMA schedule to their member sensor nodes for intra cluster communication. Gateways use slotted CSMA/CA MAC protocol to communicate with its next hop relay node.

The algorithms are tested extensively and depict the experimental results for both the routing and clustering in a combined way. For the sake of simulation, two different network scenarios (WSN#1 and WSN#2) are considered. Both of them have the sensing field of 500×500 square meter area. For the WSN#1, the position of the base station was taken at (500,250), i.e., in a side of the region and for the WSN#2, the position of the base station was taken at (250,250), i.e., in the centre of the region. To execute the algorithms, an initial population of 60 particles and the values

TABLE 6.8: PSO Parameters

Parameters	Value
N_P	60
c_1	1.4962
c_2	1.4962
w	0.7968
V_{max}	0.5
V_{min}	-0.5

of PSO parameters are taken same as in [160, 162] as shown in Table 6.8. The size of the swarm can be defined differently. However, here a pre-defined swarm size of 60 particles is used. The same experiment can be tested for 50 or 70 particles. In weight sum approach it is tested for different values of the weight factor, w_1 and w_2 and observed that for $w_1 = 0.2$ and $w_2 = 1 - w_1 = 0.8$, it was showing comparably better result. Therefore, in the simulation the same value have been taken. It should be noted that it is very difficult to precisely and accurately select these weight values, even for someone familiar with the problem domain [161].

For the sake of comparison, the GA-based clustering algorithm as presented in Section 6.3 and another two clustering algorithms GLBCA [32] and LDC [33] are executed. Note that all the above three algorithms assume that the base station is within the direct communication range of the gateways and thereby they do not consider any multi-hop routing algorithm between gateways and the base station. As

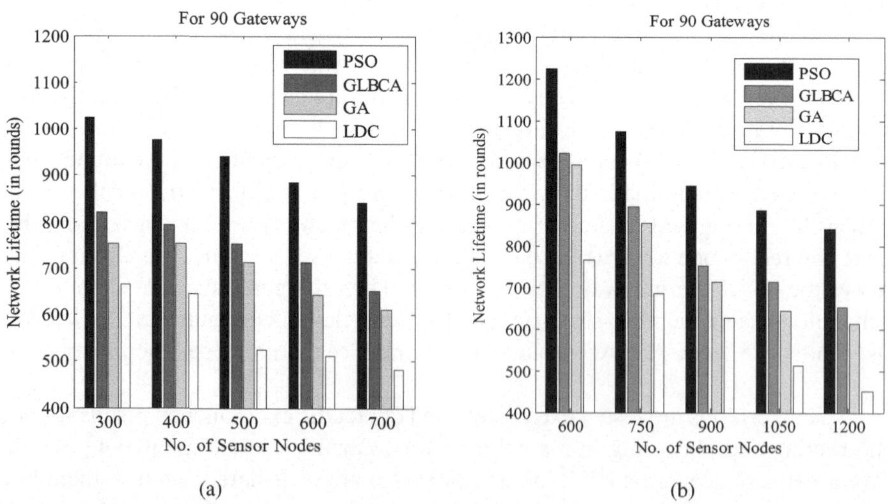

FIGURE 6.20: Comparison in terms of network life in rounds for 90 gateways in (a) WSN#1 and (b) WSN#2

TABLE 6.9: Mean network life time and standard deviation in WSN#1(60 gateways)

Algorithms	200 Sensor nodes		400 Sensor nodes		600 Sensor nodes	
	Mean	SD	Mean	SD	Mean	SD
PSO	831.23	22.21	814.12	21.21	691.25	19.36
GLBCA	715.21	23.21	661.14	23.53	601.41	20.54
GA	674.11	21.14	651.71	23.64	541.87	18.63
LDC	643.52	26.84	523.39	24.32	441.69	19.23

the work consists of clustering algorithm along with multi-hop routing, for the fair comparison a popular GA-based multi-hop routing algorithm [31] is executed for the three clustering algorithms.

First, the algorithms are compared comparing lifetime of the network by varying the sensor nodes from 200 to 700 and the number of gateways for 60 and 90 on both of the network scenarios, WSN#1 and WSN#2. Figure 6.20 shows the comparison of the PSO-based algorithm with GLBCA, GA and LDC in terms of network life in WSN#1 and WSN#2 respectively. It can be observed from Figure 6.20 that the PSO based algorithm has better network lifetime than the GLBCA, LDC and GA-based clustering algorithms. This is due to the clustering phase, where it takes care of the CHs those are inevitably used as a relay node to forward the data packets to the base station. Thus it helps to delay in initial death of the CH and increase network lifetime. Whereas the existing clustering algorithms do not deal with the uneven data forwarding effects on the CHs, thereby the CHs those are near to the base station die quickly due to extra work load of frequent data forwarding. The average (Mean) network life time for 25 runs of the algorithms along with their standard deviations (SD) for both the scenarios WSN#1 and WSN#2 are also calculated by varying number of sensor nodes and gateways. The results are shown in Table 6.9 and Table 6.10 for 60 and 90 gateways respectively. It is clear that the average network

TABLE 6.10: Mean network life time and standard deviation in WSN#2(90 gateways)

Algorithms	300 Sensor nodes		500 Sensor nodes		700 Sensor nodes	
	Mean	SD	Mean	SD	Mean	SD
PSO	1185.63	27.56	954.85	23.51	874.53	26.35
GLBCA	1036.32	29.35	774.52	24.12	675.12	24.13
GA	986.74	24.56	714.23	21.54	621.98	29.41
LDC	754.31	29.54	621.5	28.14	422.12	28.53

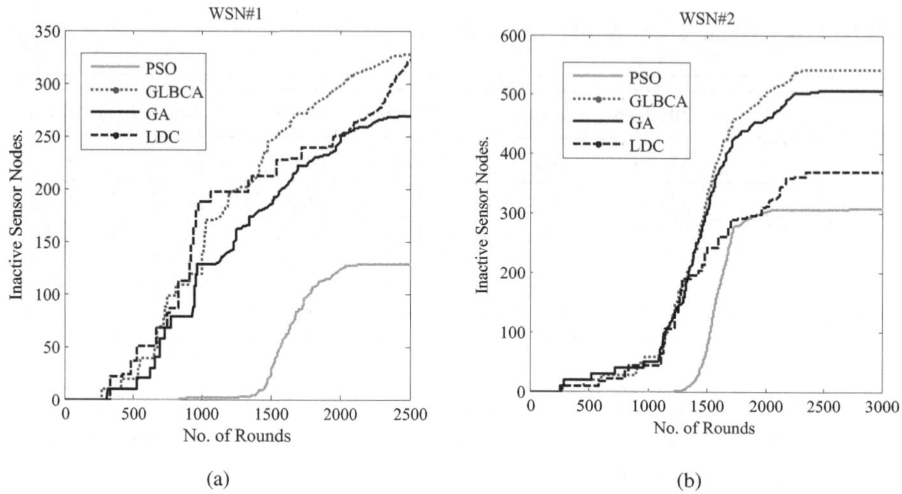

FIGURE 6.21: Comparison in terms of inactive sensor nodes in (a) WSN#1 and (b) WSN#2

life time for the PSO based algorithm is maximum. However, GA has the minimum fluctuations in the average network life.

Now, the comparison of the algorithms is shown in terms of number of inactive sensor nodes per round in WSN#1 and WSN#2 respectively. A sensor node is considered as active if its existing energy is not zero and also there must be at least one CH within its communication range. Sometimes few CHs die quickly for improper load balancing in clustering and extra overload of data forwarding. As a result, few sensor nodes are unable to find any CH within their range, though they still may have some existing energy. In the scenario this type of sensor nodes are also considered as inactive. Simulations are performed by means of different algorithms for 600 sensor nodes and 60 gateways. It can be observed from Figure 6.21 that the rate of inactive of the sensor nodes in both of the scenarios for the algorithm is lesser than the existing algorithms. This is due to the fact that the derived fitness function takes care about the energy consumption of the normal sensor nodes by reducing the distances between sensor nodes and the gateways. Moreover, long life of the CHs helps the sensor nodes to be active for the long time. It can be observed that GLBCA and the GA-based clustering algorithms only balance the load of the CHs. To achieve this goal, some sensor nodes are assigned to the CH which may be farther from it. As a result their energies are drained out due to long haul transmission and die quickly. Whereas, though LDC assigns the sensor nodes to their nearest CH to reduce energy consumption of the normal sensor nodes, it does not take care of the load balancing of the CHs and the data forwarding overhead. In a result, frequent death of the CHs lead some sensor nodes become inactive though they may have some remaining energy.

Figure 6.22 shows the comparison of energy (J) consumption of the network per

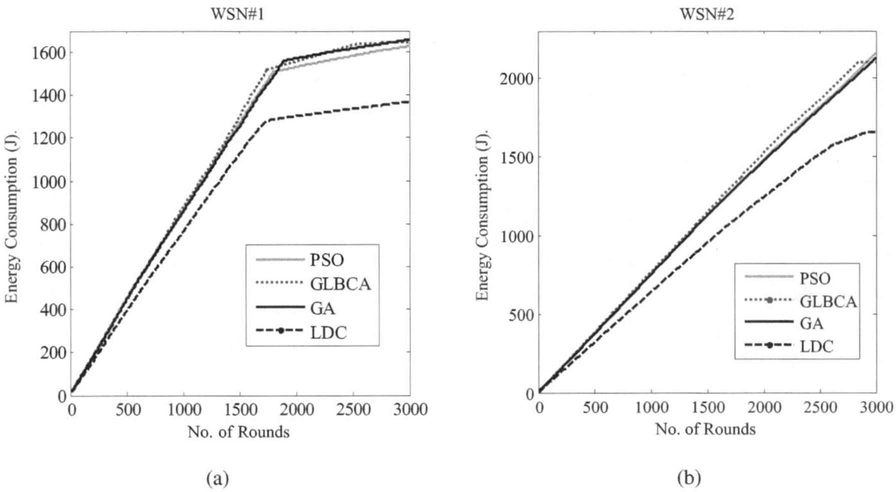

FIGURE 6.22: Comparison in terms of energy consumption in (a) WSN#1 and (b) WSN#2

round for 600 sensor nodes and 60 gateways in WSN#1 and WSN#2 respectively. Though the PSO based algorithm, GLBCA and the GA-based clustering algorithms consume more or less same amount of energy it can be claimed that the PSO based algorithm performs better in this respect. The justification behind it is that the higher number of active sensor nodes in the network consume more energy than the others. LDC consumes comparably lesser energy due to its sensor assignment strategy. As the sensor nodes are assigned to their nearest CH, they consume less energy and as a result the overall energy consumption of the network becomes lesser than the others. However, overall performance of the network is not only the measurement of energy consumption. It should be noted that in terms to assign the sensor nodes to their nearest CH, load of the CHs is not properly balanced which leads to initial death of the overloaded CHs. The initial death of the CHs may cause the network be disconnected and the maximum sensed data packets are unable to reach to the base station. It is also observed that the number of data packets received by the base station in LDC is comparably very lesser than the PSO based algorithm as well as others.

To evaluate the performance of the PSO based routing algorithm, two GA-based routing algorithms, i.e., GAR [30] and the algorithm proposed by Ataul et al.[31] are executed. Another well known algorithm, MHRM [163] is also executed. Figure 6.23 shows the comparison of the algorithms in terms of the total number of hops used. In this case, MHRM always shows the better result. This is because MHRM uses maximum possible distance for the next hop selection, thus it can easily reduce the hop count but unfortunately it extends the transmission distance and use long haul transmission. Whereas, GAR minimizes the transmission distance to reduce the energy consumption but it uses comparably high number of hops. The PSO-based

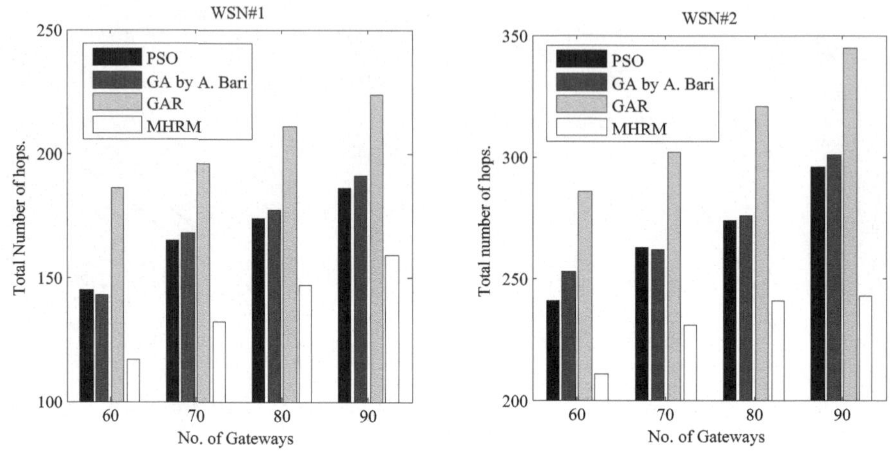

(a) (b)

FIGURE 6.23: Comparison in terms of total number of hops for (a) WSN#1 and (b) WSN#2

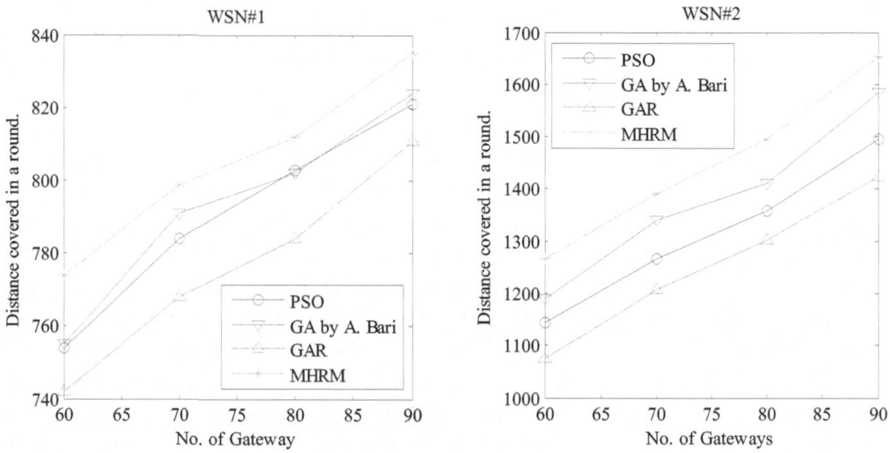

(a) (b)

FIGURE 6.24: Comparison in terms of total distance covered in a round for (a) WSN#1 and (b) WSN#2

routing algorithm builds a trade-off between transmission distance and hop count, i.e., GAR and MHRM. All the algorithms are also compare with respect to the total distance covered in each round. This can be observed from Figure 6.24 that the PSO based algorithm covers comparably less distance than MHRM and GA-based routing algorithm [31] in a single round.

6.6 Chapter summary

In this chapter, first a GA based load balanced clustering algorithm is presented for WSN. The algorithm has been described with proper chromosome representation, generation of initial population, selection process, followed by the crossover and mutation operations. The experimental results have shown that the performance of the algorithm is better than the GA based clustering algorithm, simple GA, DE, Load Balanced Clustering (LBC) and the Least Distance Clustering (LDC) algorithm in terms of load balancing of the gateways for equal as well as unequal load of the sensor nodes. It is observed that the algorithm provides superior performance in terms of energy consumption, number of active sensor nodes, rate of convergence and the execution time. However, the algorithm does not consider residual energy of the gateways and energy consumption of the sensor nodes.

Next, a DE based Clustering Algorithm for WSNs called DECA is presented. DECA takes care of the lacks of the above GA based clustering algorithm. The approach has introduced an efficient vector encoding scheme and an extra phase called local improvement to improve the DE based clustering algorithm. The experimental results have shown that the DE based algorithm converges faster than the traditional DE and GA and also performs better than the exiting algorithms in terms of network life, energy consumption and number of dead sensor nodes.

Finally, PSO based clustering and routing algorithms are also presented. The routing algorithm builds a trade-off between transmission distance and number of hop-count. To balance the energy consumption of the CHs, routing overhead of the CHs is considered in the clustering phase. The CHs those are heavily used as next hop relay in data forwarding is been assigned by lesser number of sensor nodes. Thus the rate of energy consumption of the CHs is significantly balanced and the lifetime of the network is improved. The algorithms are based on the derivation of efficient particle encoding schemes and fitness functions separately for routing and clustering. It is shown that the PSO based algorithms perform better in terms of network life, number of inactive sensor nodes and total data packets transmission.

Chapter 7

Distributed Algorithms

Chapter 7

Distributed Algorithms

7.1 Introduction

All the algorithms described in the previous chapters are centralized in nature and hence, time consuming. Let us study few distributed algorithms for both the clustering as well as routing problem.

In some applications CHs are selected from normal sensor nodes and a model of such type cluster based WSN is shown in Figure 7.1. In such type of cluster based architecture, the selected CHs die quickly due to the extra work load of head. Moreover, the CHs, which are closer to the base station or on some energy efficient path, are inevitably used as a relay node to forward the packet to the base station and thereby drain their energy very quickly. Therefore, selection of CHs with cluster setup and next-hop selection in routing need to be properly addressed for balancing the energy consumption of the sensor nodes to improve the network lifetime.

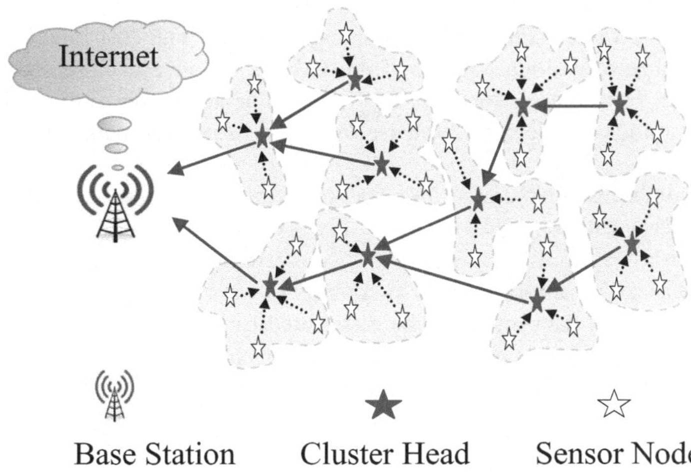

FIGURE 7.1: A WSN model with normal sensor nodes as CHs

In the data routing phase, CHs send the processed data through other CHs in which intermediate CHs act as relay nodes. But it requires the guarantee of connectivity among the CHs along the routing path. This is worth noting that the intermediate non-CH sensor nodes may be used as relay nodes between two unconnected CHs, but it may require special algorithmic design and also generate extra control message overhead on the network which may lead to serious energy inefficiency. Nevertheless, there may not be always a sensor node between two unconnected CHs which can act as a relay node. Therefore, proper care should be taken in the selection of CHs, especially whenever CHs are selected from the normal sensor nodes. Moreover, the routing algorithm should take care the energy consumption of the nodes to prolong the network lifetime. In order to be energy efficient, the algorithm is supposed to route the sensed data via other sensor nodes which are closer to the base station or on some energy efficient path. But this arrangement of data routing can lead the sensor nodes of the energy efficient path die quickly for frequently used as relay node. This uneven energy consumption decreases the coverage ratio and reduces the network lifetime. If the sensor nodes consume their energy more evenly, the connectivity between them and the BS can be maintained for a longer time and thus it prevents the network to be partitioned. Therefore, it is very important to make an appropriate trade-off between energy efficiency and balanced energy consumption [117].

In this chapter, first a cluster based routing algorithm called CBCRA (Cost Based Clustering and Routing Algorithm) is presented. It is a distributed algorithm which consists of three phases namely selection of CHs, cluster setup and data routing. The algorithm is fully based on the local information of a sensor node such as residual energy, number of neighbors and their distances. CBCRA selects CHs amongst the normal sensor nodes using a weight function of the residual energy and the number of neighbors of a sensor node. Then all non-CH sensor nodes join a CH, which has maximum cost value within its communication range. The cost function is the composite measure of residual energy of the CH, its distance to base station and also the distance from the sensor node to the CH. For multi-hop routing, the best neighbour relay node is selected by measuring the cost of each path towards base station.

Second, another distributed algorithm for clustering sensor nodes called DEECR (Distributed Energy Efficient Clustering & Routing) is also presented. Here, CHs are selected in such a way that it guarantees the connectivity between the CHs during data routing phase. The algorithm is also consists of three phases, i.e., selection of CHs, cluster setup and data routing. The distributed routing algorithm selects next hop relay nodes by fulfilling two necessary conditions based on the communication distance among the consecutive CHs and the base station.

7.2 Network model and terminologies

In this assumed model, all the sensor nodes are homogeneous , i.e., they have the same processing capabilities and also have the same communication and broadcasting power level. Here, CHs are selected from the normal sensor nodes and a model of such type cluster based WSN is shown in Figure 7.1. Initially, all the sensor nodes are charged with the same amount of energy. Sensor nodes and the base station are all stationary after deployment. Sensor nodes can use power control to vary the amount of transmission power which depends on their distance between sender and the receiver like Berkeley Motes [164]. All communication is over wireless links. A wireless link is established between two nodes only if they are within the communication range of each other. The wireless links are symmetric in this model. Given the transmitting power, a node can compute the approximate distance to another node based on the received signal strength as in [103, 104, 165, 166, 167, 168].

Here, the terminologies and notations are used same as used in the previous chapters. However, a few extra terminologies and notations are used for the presentation of the algorithm defined as follows.

- The set of cluster heads (CHs) selected from the normal sensor nodes is denoted by $C = \{C_1, C_2, C_3, \ldots\}$. Note that, $C \subseteq S$.

- $ComRangeSet(s_i)$ is the set of all those nodes, which are within the communication range (R_S) of node s_i. Therefore,

$$ComRangeSet(s_i) = \{s_j | dis(s_i, s_j) \leq R_S \wedge s_j \in \{S - s_i\}\} \tag{7.1}$$

- $Neighbor(s_i)$ is the set of all those nodes, which are within the cluster range (R_C) of node s_i. Therefore,

$$Neighbor(s_i) = \{s_j | dis(s_i, s_j) \leq R_C \wedge s_j \in \{S - s_i\}\} \tag{7.2}$$

It should be noted that in the assumed scenario, cluster range is always less than communication range. It means

$$R_C < R_S \text{ and } Neighbor(s_i) \subseteq ComRangeSet(s_i) \tag{7.3}$$

- $AvgEnergy(s_i)$ denotes the average energy of the sensor node s_i itself and its neighbor. Therefore,

$$AvgEnergy(s_i) = \frac{\sum E_{residual}(s_j), \forall s_j \in \{Neighbor(s_i) \cup s_i\}}{|Neighbor(s_i)| + 1} \tag{7.4}$$

- $CH\_Neighbor(C_i)$ denotes the set of all CHs within the communication range of C_i which are towards the BS. Therefore,

$$CH\_Neighbor(C_i) \tag{7.5}$$
$$= \{C_j | C_j \in ComRangeSet(C_i) \wedge dis(C_i, BS) > dis(C_j, BS) \wedge i \neq j\} \tag{7.6}$$

- $H_C(s_i)$ denotes the number of next hops required to reach to the BS from s_i. If s_i can directly communicate with BS, then $H_C(s_i)$ is one. Therefore, $H_C(s_i)$ can be recursively defined as

$$H_C(s_i) = \begin{cases} 1 & \text{If } dis(s_i, s_{N+1}) \leq d_0 \\ 1 + H_C(s_j) & H_C(s_j) = \text{Min}\{H_C(s_k) | \\ & dis(s_i, s_k) \leq d_0, \forall s_k\} \end{cases} \tag{7.7}$$

7.3 Distributed cost based algorithm

At the beginning, all the sensor nodes undergo bootstrapping process, in which they are assigned unique IDs. The BS broadcasts a HELLO message to all the sensor nodes at a certain power level. By this way, each node can compute the approximate distance to the BS based on the received signal strength. It helps the sensor nodes to select the proper power level to communicate with the BS. This approach is also used in many algorithms such as presented in [103, 104, 165, 166, 167, 168]

Network setup consists of a setup phase followed by the steady-state phase as shown in Figure 7.2 and illustrated as follows. Setup phase performs selection of the cluster heads and forms clusters using the distributed algorithm discussed later. Steady-state phase divides the schedule of the network into multiple rounds of fixed duration. In each round, the cluster heads receive the sensed data from the cluster member nodes and aggregate them to transfer it to the base station using the distributed multi-hop routing algorithm discussed later. Now the details of cluster head selection, cluster formation and inter cluster routing are presented in the following sections subsequently.

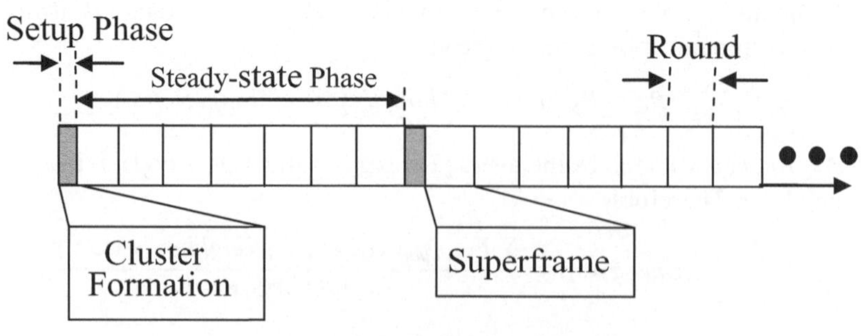

FIGURE 7.2: Timing in our Protocol

7.3.1 Clustering algorithms

7.3.1.1 Selection of CHs

The method exploits the localized network structure and the remaining energy of the sensor nodes for selection of the cluster heads. We use a weight function $W(s_i)$ which depends on various parameters described as follows [169].

1. ***Residual Energy*:** CHs consume more energy than any ordinary sensor node. We have to select those sensor nodes as CHs which can perform more rounds as a CH. Higher the residual energy more is the chance to becoming a CH. This implies that
$$W(s_i) \propto E_{residual}(s_i) \tag{7.8}$$

2. ***Ratio of Residual Energy and Neighbor*:** The sensor node, which is selected as a CH, has to perform more rounds than simple sensor nodes i.e., the average amount of energy of the selected sensor node should be higher than that of any of its neighbour sensor nodes. It means
$$W(s_i) \propto \frac{E_{residual}(s_i)}{Neighbor(s_i)} \tag{7.9}$$

Condition (7.8) and condition (7.9) combinedly implies that

$$W(s_i) \propto E_{residual}(s_i) \times \frac{E_{residual}(s_i)}{Neighbor(s_i)}$$

$$i.e., \quad W(s_i) = K_1 \times E_{residual}(s_i) \times \frac{E_{residual}(s_i)}{Neighbor(s_i)}$$

where K_1 is a proportionality constant. It is noteworthy that we take the weight value of the sensor nodes only for comparison purpose. So, the value of K_1 does not hamper our objective and without any loss of generality, we can assume $K_1 = 1$. Therefore,

$$W(s_i) = E_{residual}(s_i) \times \frac{E_{residual}(s_i)}{Neighbor(s_i)} \tag{7.10}$$

The idea behind the CH selection is as follows. First of all, each sensor node broadcasts a HELLO message within its communication range (R_S). The HELLO message consists of node-id and residual energy of the sensor node. Now sensor node can estimate its number of neighbors by counting the received HELLO messages. Then they calculate their weight value using equation (7.10) and broadcast it within its cluster range. If a sensor node receives higher weight value from any of its neighbors then it gives up the CH selection competition and become inactive to save energy. Otherwise, it declares itself as a CH and broadcasts an advertisement message using CSMA/CA MAC [135, 168] protocol. Any sensor node which does not receive any CH advertisement after a certain timeout, declares itself as a CH. A sensor node declares itself as a CH only if it has not received any CH advertisement.

7.3.1.2 Cluster setup

After CH declaration, each non-CH sensor node joins a CH based on the value of the cost function described as follows. Let $CH\_Cost(C_i, s_j)$ be the cost of the cluster head C_i for the sensor node s_j. To define the cost function we consider the following parameters.

1. ***Residual Energy of CH*:** Sensor node should join that CH which has higher residual energy than any other CH within its cluster range. Therefore,

$$CH\_Cost(C_i, s_j) \propto E_{residual}(C_i) \tag{7.11}$$

2. ***Distance from Sensor node to CH*:** As non-CH sensor nodes consume maximum energy to communicate with its CH, sensor node should join the nearest CH. The shorter the distance higher is the chance to join. Therefore,

$$CH\_Cost(C_i, s_j) \propto \frac{1}{dis(C_i, s_j)} \tag{7.12}$$

3. ***Distance from CH to Base Station*:** The CHs nearer to the BS, consumes more energy inevitably as they frequently act as the relay node in routing phase. Therefore, cluster member of these CHs should be less than that of the CHs which are far away from the BS. In other words,

$$CH\_Cost(C_i, s_j) \propto dis(C_i, BS) \tag{7.13}$$

By combining equations (7.11), (7.12) and (7.13), we obtain

$$CH\_Cost(C_i, s_j) \propto \frac{E_{residual}(C_i) \times dis(C_i, BS)}{dis(C_i, s_j)}$$

$$i.e., \quad CH\_Cost(C_i, s_j) = K_2 \times \frac{E_{residual}(C_i) \times dis(C_i, BS)}{dis(C_i, s_j)}$$

where, K_2 is proportionality constant. In the same way we can assume K_2 =1. Therefore,

$$CH\_Cost(C_i, s_j) = \frac{E_{residual}(C_i) \times dis(C_i, BS)}{dis(C_i, s_j)} \tag{7.14}$$

Now s_j selects the cluster head say C_k with highest cost value. Therefore,

$$CH\_Cost(C_k, s_j) = \text{Max}\{CH\_Cost(C_i, s_j) \mid \forall C_i \in Neighbor(s_j)\} \tag{7.15}$$

Each non-CH sensor node selects a CH to join using equation (7.15) and then sends a JOIN_REQ message to its selected CH using CSMA/CA MAC protocol. This ends the cluster setup phase.

All cluster members use single-hop communication to communicate with their CH. For such intra-cluster communication, all the CHs set up a superframe as per

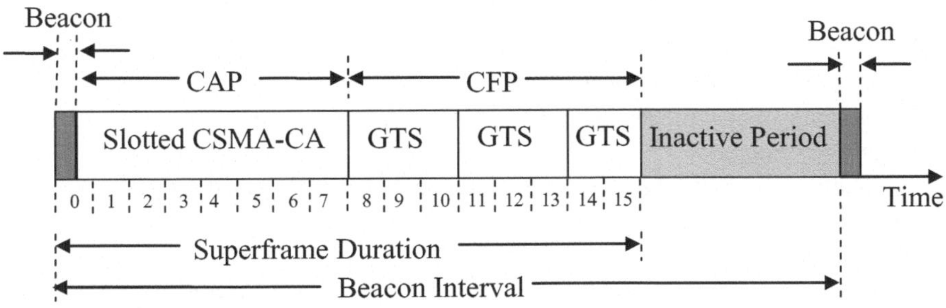

FIGURE 7.3: The superframe structure of the IEEE 802.15.4 beacon-enabled mode

IEEE 802.15.4 MAC which supports beacon enable superframe-based operating mode [11, 135, 168]. The scheme is illustrated in Figure 7.3 and operated as follows. The CHs start the superframe with a beacon. The beacon is used to synchronize communication and to convey the superframe structure to the cluster members. The superframe is divided into an active and an inactive period. In inactive period, the cluster heads may go to sleep mode and save their energy. The active period is divided into fixed size slots containing a contention access period (CAP) and contention free period (CFP). In CAP, sensor nodes compete for channel access using a slotted CSMA/CA protocol. In CFP, cluster heads assign guaranteed time slots (GTS) to the member sensor nodes for the channel utilization. A non-CH sensor node willing to communicate with its CH (in CAP), it must wait for the beacon to synchronize and then compete for channel access. Communication from a CH to a member sensor node is as follows. The CH stores the request message and announces pending delivery in the beacon. The non-CH sensor nodes usually go to sleep mode and wake up periodically to check for some message from the CH in the next beacon. When they notice that a message is available from CH, they communicate with the CH during CAP. When a CH wishes to communicate with another CH, it must synchronize with the beacon of other CH and act as a member sensor node. The duration of a round is same as beacon interval as shown in Figure 7.2.

The algorithm for CH selection and cluster formation for sensor node s_i is formalized in Algorithm 7. The algorithm is simultaneously run by each sensor node $s_i, \forall i, 1 \leq i \leq n$, for cluster formation. This is obvious to note that the algorithm select each sensor node to act as either a CH or a cluster member.

Lemma 7.1. *The CH-Select-Cluster-Formation algorithm has worst case time complexity of $O(n)$ per sensor node, where n is the number of sensor nodes in the network.*

Proof. In the CH selection phase, a sensor node has to compare its weight to the neighbor nodes to declare itself as a CH. For a network with n number of sensor nodes, the possible number of neighbor is $n - 1$ in the worst case. So, the CH selection phase takes $O(n)$ processing time for each node. Now, all the non-CH sensor nodes join the CH with highest cost value. In worst case, it takes $O(n)$ processing

Algorithm 7 CH-Select-Cluster-Formation

Input: Residual energy i.e., $E_{residual}(s_i)$.
Output: s_i as a CH or a cluster member.

s_i broadcasts and receive HELLO message to and from all the sensor nodes within RS range.
s_i counts $Neighbor(s_i)$ according to the received HELLO message and calculates $W(s_i)$ using equation (7.10).
flag=1.
s_i senses the channel to broadcasts $W(s_i)$ within RS range and receives $W(s_j)$ from neighbor.
if $(W(s_j) > W(s_i))$ **then**
    s_i gives up CH selection competition and stays inactive.
    flag=0.
end if
if $(flag == 1)$ **then**
    s_i declares itself as a CH and broadcasts this message to all of its neighbors.
else
    Wait for advertisement messages from other nodes who have declared themselves as CH.
    if (no advertisement receive) **then**
        s_i declares itself as a CH.
    else
        Calculate the cost value of all the CHs using equation ((7.14)).
        Joins the CH with the highest cost value.
    end if
end if
Stop.

time to choose the appropriate CH. Therefore, the overall processing time of the CH selection and cluster formation algorithm is $O(n)$ + $O(n)$, i.e., $O(n)$ in worst case situation. □

Lemma 7.2. *The CH-Select-Cluster-Formation algorithm has worst case message exchange complexity of* $O(1)$ *per sensor node, i.e.,* $O(n)$ *for the whole network.*

Proof. In the CH selection phase, sensor nodes have to broadcast a HELLO message to count the number of neighbors. They need to broadcast their weight value. Then, the sensor nodes declare themselves as a CH only if they do not find any higher weighted sensor node. So, during the CH selection, a sensor node has to broadcast at most 3 messages in worst case. In cluster formation phase, all non-CH sensor nodes have to send the join message to the selected CH. Therefore, overall message exchange complexity of a sensor node is constant, i.e., $O(1)$ and $O(n)$ for the whole network. □

7.3.2 Routing algorithm

We now outline here the scheme for data aggregation and routing. At the end of cluster formation as described in previous section, all sensor nodes sense the local data and send it to its CH. Once the CHs receive the data from all their corresponding members, they perform data aggregation to reduce the redundant and uncorrelated data within their cluster in distributed manner. This completes data aggregation.

After data aggregation, CHs route their aggregated data either directly to BS or via other CHs treating as relay node. The CHs determine another CH as a best candidate relay node to communicate with the BS among its neighbors and itself. The CH selects the relay node in such a way that it pursues energy balance for the sensor network. This is implemented as follows.

It is assumed that, each CH also aware of the remaining energy and distance of all CHs within its communication range from the BS [168]. To select a proper relay node, the method calculates cost of each path. The path cost indicates nothing but the suitability of a path to achieve the energy balance of the network. If a cluster head C_i sends the aggregated data to base station via a relay node C_r then the total path cost $PCost(C_i, C_r)$ depends on few factors as follows

1. ***Transmission Energy*:** The next hop relay node C_r should be selected such a way that it requires minimum energy to transmit the data from C_i to C_r. Therefore,
$$PCost(C_i, C_r) \propto \frac{1}{E_{Transmit}(C_i, C_r)} \tag{7.16}$$

2. ***Residual Energy*:** To be selected as a next hop relay node, C_r must have sufficient available energy, i.e.,
$$PCost(C_i, C_r) \propto E_{residual}(C_r) \tag{7.17}$$

3. ***Required Transmission Energy of C_r*:** The relay node C_r receives the data from sender C_i and transmits the data to base station. So, total required transmitting energy and receiving energy is one of the most important factors. Obviously, the less required total energy indicates a suitable path. In other words
$$PCost(C_i, C_r) \propto \frac{1}{E_{Transmit}(C_r, BS) + E_R} \tag{7.18}$$
where, $E_{Transmit}(C_r, BS)$ is the required energy to transmit data from C_r to BS and E_R denotes the required energy to receive the packet.

The condition (7.16), condition (7.17) and condition (7.18) combinedly implies that

$$PCost(C_i, C_r) \propto \frac{E_{residual}(C_r)}{E_{Transmit}(C_i, C_r) \times \{E_{Transmit}(C_r, BS) + E_R\}}$$

$$i.e., \qquad PCost(C_i, C_r) = K_3 \times \frac{E_{residual}(C_r)}{E_{Transmit}(C_i, C_r) \times \{E_{Transmit}(C_r, BS) + E_R\}}$$

where, K_3 is proportionality constant. Without loss of generality, we have assume that $K_3 = 1$. Therefore,

$$PCost(C_i, C_r) = \frac{E_{residual}(C_r)}{E_{Transmit}(C_i, C_r) \times \{E_{Transmit}(C_r, BS) + E_R\}} \quad (7.19)$$

If, C_i transmits the data directly to BS then the path cost is calculated as

$$PCost(C_i, BS) = \frac{1}{E_{Transmit}(C_i, BS)} \quad (7.20)$$

The basic idea of the method is to select a path having maximum cost. When a CH sends data to the base station, it first selects a best candidate from its neighbour and itself considering the maximum cost of the path following the equation (7.19) or (7.20). If, the CH itself is a best candidate, then the CH itself sends the data to BS otherwise, it forwards the data to the selected best candidate node. Then, the selected best candidate node repeats the same routing process until a CH selects itself as a best candidate and sends the data directly to base station. The relay node selection algorithm for cluster head C_i is given in Algorithm 8.

The experiments are performed with 150 sensor nodes placed in a 200 × 200 square meter area and the position of the sink is (175, 225). The experimental results are shown in Figure 7.4. It is observed from Figure 7.4(a) that in the case of DEBR and LEACH, almost all sensor nodes are inactive within 1600 rounds and 2700 rounds respectively whereas, in the case of CBCRA, they take 3300 rounds to be inactive. It is also shown in Figure 7.4(b), that the energy consumption of CBCRA is drastically lower than DEBR as well as LEACH.

Algorithm 8 Next-Hop-Select

Input: 1) Residual energy of all CHs from *Neighbour*(C_i) and 2) Distance from C_i to BS and all CHs from *Neighbour*(C_i).
Output: Next-hop relay node (*Relay*) of C_i.

```
Relay = BS /* initially, relay node is BS */
T = Neighbour(C_i)
while (T ≠ NULL ) do
    Select C_r from T.
    if (PCost(C_i, C_r) > PCost(C_i, Relay)) then
        Relay = C_r
    end if
    T = T − C_r
end while
Stop.
```

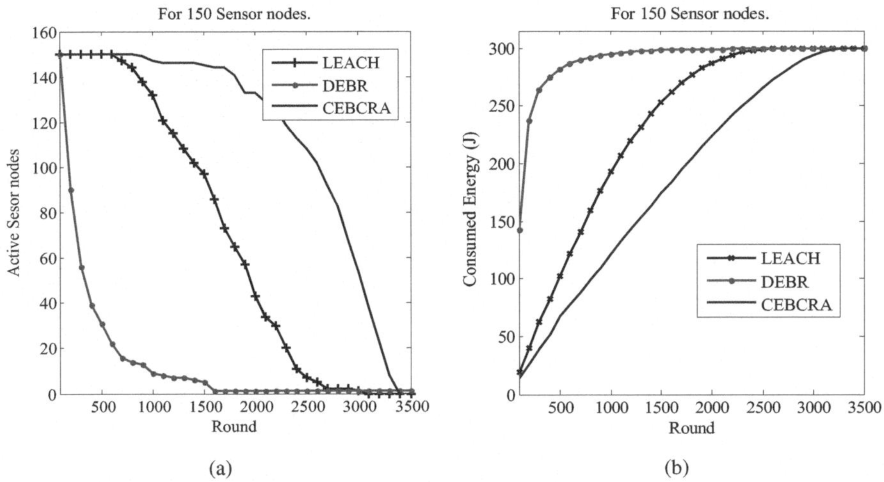

FIGURE 7.4: Comparison between LEACH, DEBR and CBCRA in terms of (a) active sensor nodes and (b)energy consumption (J) for 150 sensor nodes

7.4 Distributed energy efficient algorithm

Here, initial network setup is same as it is used in the previous algorithm as presented in section 7.3. Note that the previous algorithm does not assured the connectivity between selected CHs. Now, cluster size is chosen in such a way that it guarantees the connectivity amongst CHs during the routing phase. This is illustrated as follows.

Let us consider a worst case situation in which cluster range is taken maximal so that the CHs can just communicate with each other. Figure 7.5 presents such a scenario in which the cluster range (R_C) is shown by dashed circle and the communication range (R_S) by solid circle (bigger one). In this figure $C_1, C_2, C_3, \ldots, C_6$ are the selected CHs which are just within their communication range. This indicates that the distances C_1C_2, C_1C_3, C_1C_7, C_2C_3, C_2C_4, C_3C_4, C_3C_5, C_3C_6, C_3C_7, C_4C_5, C_5C_6 and C_6C_7 are all equal and they represent the communication range of the CHs. This should be noted that for any cluster range greater than (R_C), the clusters may be formed in such a way that the connectivity among the CHs cannot be guaranteed.

Lemma 7.3. *The connectivity among the CHs is guaranteed if* $R_C \leq \frac{R_S}{\sqrt{3}}$

Proof. As the above scenario (refer Figure 7.5) is a worst case situation, the areas are minimally covered by the CHs. It implies that the cluster perimeters of C_1, C_2 and C_3 meet at the same point say B which is the centroid of $\triangle C_1C_2C_3$ as it is an equilateral triangle.

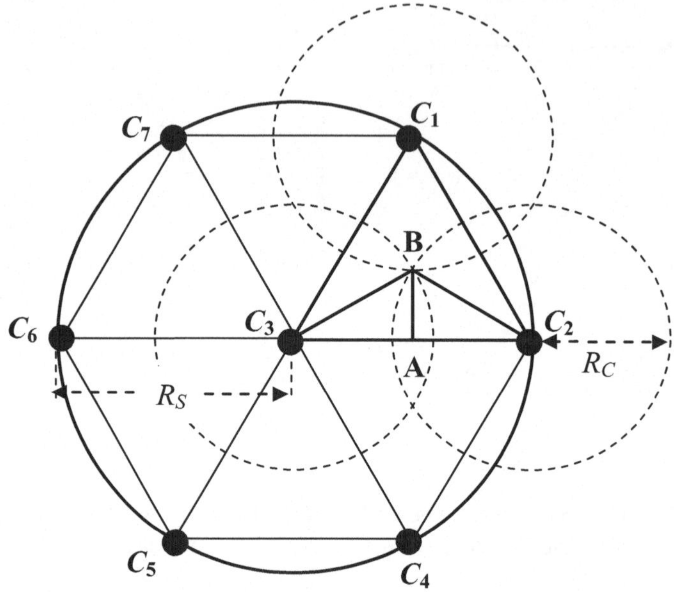

FIGURE 7.5: Cluster size calculation

Let, A be in the middle of C_2C_3. Then BA is perpendicular to C_2C_3. Now,

$$\angle AC_3B = \frac{\Pi}{6}$$

$$\therefore \frac{C_3A}{C_3B} = \cos \angle AC_3B = \frac{\sqrt{3}}{2}$$

$$\text{i.e., } \frac{\frac{R_S}{2}}{R_C} = \frac{\sqrt{3}}{2}$$

$$\text{i.e., } R_C = \frac{R_S}{\sqrt{3}}$$

Therefore, the connectivity is assured only if $R_C \leq \frac{R_S}{\sqrt{3}}$ □

7.4.1 Clustering algorithms

7.4.1.1 Selection of CHs

Here, also we use the weight function $W(s_i)$ to select the CHs. However, except the residual energy of the sensor nodes we use two new parameters described as follows.

1. ***Residual Energy***: As it is previously described, the weight value is propor-

tional to the residual energy.

$$W(s_i) \propto E_{residual}(s_i) \tag{7.21}$$

2. ***Number of Neighbors*****:** Selected CH should have sufficient neighbors to become cluster member. So, a sensor node having higher number of neighbors should have more chance to become a CH. In other words,

$$W(s_i) \propto Neighbor(s_i) \tag{7.22}$$

3. ***Ratio of Residual Energy and Average Energy*****:** The above two factors may cause a sensor node to become a CH without having sufficient residual energy but with many number of neighbors. To overcome this situation, we introduce the third factor, which is the ratio of residual energy and the average energy of the sensor nodes. If the sensor node has less energy, then the ratio is less than 1, otherwise it is greater than 1. Thus the sensor node which has less residual energy but many number of neighbors has lower chance to become a CH. So, higher the ratio more is the chance to becoming a CH. Therefore,

$$W(s_i) \propto \frac{E_{residual}(s_i)}{AvegEnergy(s_i)} \tag{7.23}$$

Now we combine all the conditions (7.21), (7.22) and (7.23) to obtain

$$W(s_i) \propto E_{residual}(s_i) \times Neighbor(s_i) \times \frac{E_{residual}(s_i)}{AvegEnergy(s_i)}$$

$$i.e., \qquad W(s_i) = K_4 \times E_{residual}(s_i) \times Neighbor(s_i) \times \frac{E_{residual}(s_i)}{AvegEnergy(s_i)}$$

where K_4 is a proportionality constant. In the same way we can assume $K_4 = 1$. Therefore,

$$W(s_i) = E_{residual}(s_i) \times Neighbor(s_i) \times \frac{E_{residual}(s_i)}{AvegEnergy(s_i)} \tag{7.24}$$

The CH selection process is same as discussed in the previous section. However, few changes are as follows. The sensor node broadcasts the HELLO message within its cluster range (R_C) in contrast to communication range (R_S) as it is used in the CBCRA. Note that it helps to select the CHs in such away that the connectivity can be assured as discussed.

7.4.1.2 Cluster setup

Here, same cluster setup is used as described in section 7.3.1.2.

7.4.2 Routing algorithms

In this algorithm, CHs select the relay node in such a way that it requires minimum energy to transmit the aggregated data to the BS. It also avoids the relay nodes with less energy that may fall in a routing path to BS. This is implemented as follows. Every CH is aware about the distance from its neighbor CHs, its own distance from the BS and also its neighbor CH distance from the BS as in [104, 135, 168]. A CH may have many possible CHs as relay node towards BS. To efficiently balance energy consumption, a CH selects some eligible CHs from these possible CHs. A CH is eligible to be a relay node if its energy level is above some threshold value. This algorithm chooses the threshold value as the average energy of the possible CHs. It should be noted that the threshold criteria can be modified depending on the application of this algorithm. A cluster head C_i chooses the next hop relay node C_r from the eligible nodes, only if it satisfies the following conditions.

$$(dis(C_i, B))^2 > (dis(C_i, C_r))^2 + (dis(C_r, B))^2 \tag{7.25}$$

$$D(C_i, C_r) = \text{Min}\{D(C_i, C_j) \mid \forall C_j \in \{C_{eligible} - C_i\}\} \tag{7.26}$$

where $D(C_i, C_j) = dis(C_i, C_j) \times \{(dis(C_i, C_j))^2 + (dis(C_j, B))^2\}$, B is the base station and $C_{eligible}$ is the set of eligible relay nodes.

Lemma 7.4. *A cluster head C_i chooses the next hop relay node C_r so that energy consumption is minimized if $(dis(C_i, B))^2 > (dis(C_i, C_r))^2 + (dis(C_r, B))^2$ and $dis(C_i, C_r) \times \{(dis(C_i, C_r))^2 + (dis(C_r, B))^2\}$ is minimum.*

Proof. According to equation (1.1) energy consumption depends on d^2 assuming free space communication. If a cluster head C_i sends the aggregated data to base station (i.e., B as shown in Figure 7.6) via a relay node C_r then the total energy consumption is given by

$$E_{relay} = 2lE_{elec} + l\epsilon_{fs}(dis(C_i, C_r))^2 + l\epsilon_{fs}(dis(B, C_r))^2 \tag{7.27}$$

If C_i transmits the data directly to base station, then the total energy consumption

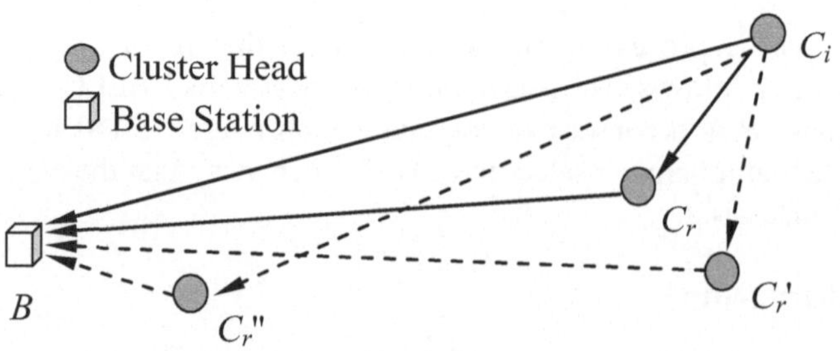

FIGURE 7.6: Selection of relay node for data transmission

Algorithm 9 Select-Relay-Node (SRN)

Input: $\forall C_j \in CH\_Neighbor(C_i)$, residual energy and approximate distance to BS.
Output: Selection of a relay node *Relay*, such that the required energy to transmit the aggregated data to BS is minimized.

```
Relay = BS /* initially, relay node is BS */
T = CH_Neighbour(C_i)
while (T ≠ NULL ) do
    Select C_r from T.
    if (C_r is eligible and satisfying conditions (7.25) and (7.26)) then
        Relay = C_r
    end if
    T = T − C_r
end while
Stop.
```

is given by

$$E_{direct} = lE_{elec} + l\epsilon_{fs}(dis(C_i, B))^2 \tag{7.28}$$

However, the cluster head C_i decides to send the aggregated data to the base station via the relay node C_r only if $E_{relay} \leq E_{direct}$, i.e.

$$\begin{aligned} & lE_{elec} + l\epsilon_{fs}(dis(C_i, B))^2 \geq 2lE_{elec} + l\epsilon_{fs}(dis(C_i, C_r))^2 + l\epsilon_{fs}(dis(B, C_r))^2 \\ or, \quad & l\epsilon_{fs}(dis(C_i, B))^2 \geq lE_{elec} + l\epsilon_{fs}(dis(C_i, C_r))^2 + l\epsilon_{fs}(dis(B, C_r))^2 \\ or, \quad & (dis(C_i, B))^2 > (dis(C_i, C_r))^2 + (dis(C_r, B))^2 \end{aligned}$$

The above derived condition ensures the existence of a relay node C_r in between C_i and B. However, there may be more than one eligible relay nodes surrounding C_i e.g., C_r'' and C_r' (as shown in Figure 7.6) which can satisfy the condition; but C_i chooses the relay node C_r such that

- C_r is closer to C_i, so that C_i need lesser energy to communicate with Cr and
- The energy consumption factor, i.e., $(dis(C_i, C_r))^2 + (dis(C_r, B))^2$ is lesser.

which jointly implies that $dis(C_i, C_r) \times \{(dis(C_i, C_r))^2 + (dis(C_r, B))^2\} = D(C_i, C_r)$ is least. □

Therefore, to transmit the aggregated data to B, a CH selects a relay node which satisfies both the conditions (7.25) and (7.26). Using same process all CHs transmit its aggregated data to B by selecting a CH as a relay node. The relay node selection algorithm for the CH C_i is given in Algorithm 9.

Lemma 7.5. *The Select-Relay-Node (SRN) algorithm guarantees the selection of a relay node among the eligible CHs such that the energy consumption for data routing to the sink via the selected relay node is minimized.*

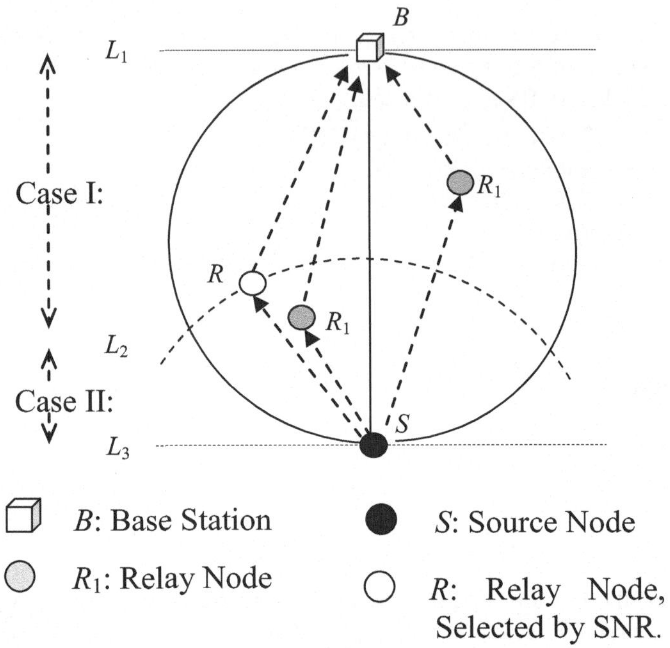

FIGURE 7.7: Relay node between source S and base station B

Proof. : Let SRN chooses R as a relay node to transmit aggregated data from source (CH) node S to B (base station) as in Figure 7.7. The SRN selects only that relay node which satisfies both of the conditions (7.25) and (7.26). For any relay node R lying on the circumference of the circle having diameter SB (as shown in Figure 7.7) fulfils $|SB|^2 = |RB|^2 + |RS|^2$ as per the Pythagoras' Theorem. Therefore, for any eligible relay node lying inside the circle follows condition (7.25) i.e., $|SB|^2 > |RB|^2 + |RS|^2$. Let there exists a better relay node say R_1 such that

$$(dis(S,R_1))^2 + (dis(B,R_1))^2 < (dis(S,R))^2 + (dis(B,R))^2 \tag{7.29}$$

The R_1 must be inside the circle. There are two possible places, where R_1 may exist as follow.

Case I: R_1 lies in between the horizontal line L_1 passing the B and L_2, the circumference of the circle assumed S as a centre and SR as radius (see Figure 7.7). Then R would be the proper relay node; because in this case R can select R_1 as next relay node to transmit the aggregated data to B and energy consumption will be less. Otherwise overall consumed energy using R_1 will be higher than using R. So, there are no R1 in between L_1 and L_2.

Case II: R_1 lies between L_2 and the horizontal line L_3, passing through S. So,

$$dis(S,R_1) < dis(S,R) \tag{7.30}$$

Thereby multiplying both the sides of equation (7.29) and (7.30), we obtain

$$dis(S,R_1) \times \{(dis(S,R_1))^2 + (dis(B,R_1))^2\} < dis(S,R) \times \{(dis(S,R))^2 + (dis(B,R))^2\} \tag{7.31}$$

But according to condition (7.26), SRN selects the relay node R which has minimum $D(S,R)$ than all other existing node inside the circle. In other words,$D(S,R_1) > D(S,R)$, i.e.,

$$dis(S,R_1) \times \{(dis(S,R_1))^2 + (dis(B,R_1))^2\} > dis(S,R) \times \{(dis(S,R))^2 + (dis(B,R))^2\} \tag{7.32}$$

Which contradicts (7.31). It means that there cannot be any relay node R_1 between L_2 and L_3. □

Lemma 7.6. *The DEECR is a fully distributed algorithm.*

Proof. :All sensor nodes execute CH selection and cluster formation algorithm depending on its local information only. In CH selection, each sensor node uses its own residual energy and number of neighbors to calculate its own weight and broadcasts the weight to decide itself, whether it is a CH or not. The CH selection process is executed by all sensor nodes simultaneously without depending on other sensor node and any global information. After CH selection, all non-CH sensor nodes select the proper CH to join as a cluster member based on the cost value of the CHs. The cost value requires local information like the energy level of the CHs which are within cluster range of the sensor node, distances from sensor node to CHs and CHs to base station. In routing phase, a CH selects other CH as a relay node using two predefined conditions. Thus DEECR is a distributed algorithm. □

Here, variation of the number of selected CHs in LEACH[19], CBCRA and DEECR is shown. It can be observed from Figure 7.8(a)-(b) that the number of selected CHs is greatly fluctuated from round to round for LEACH. In a particular round the number of selected CHs may be 60 and in very next round it may be below 5 or even 0 for 200 number of sensor nodes. This is due to the CH selection process of LEACH, which is following probabilistic approach. However, DEECR shows consistency in the number of selected CHs and it also shows the better consistency than CBCRA.

Here, the algorithms are simulated for 100 and 400 sensor nodes. Figure 7.9(a)-(b) shows the comparison results for number of active sensor nodes per round.. A sensor node is considered as active if its existing energy is not zero. It can be observed that DEECR outperforms the LEACH, DEBR [104] and CBCRA too in terms of number of active sensor nodes. DEECR performs better due to the proper CH selection and cluster formation to balance the load of the CHs and thus prevents the initial death of the CHs and non-CH sensor nodes. As, DEBR is a classical routing algorithm, it performs very poorly than other three. This is due to the fact that in DEBR all sensor nodes have to be in active mode for maximum time to collect the neighbor status which is very essential in data routing to the BS [104]. Thus sensor nodes consume more energy in idle listening and overhearing [4] and die quickly.

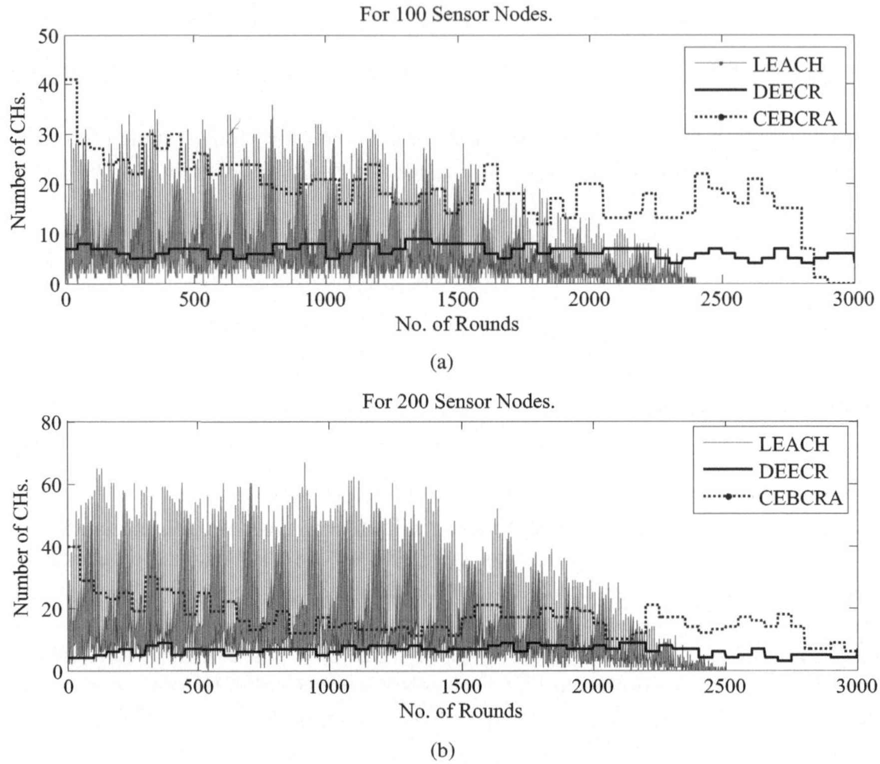

FIGURE 7.8: Comparison between LEACH, DEECR and CBCRA in terms of number of selected CHs for (a) 100 and (b) 200 sensor nodes

In LEACH, all the CHs directly send the aggregated data to BS. Therefore, CHs consume more energy due to long haul communication and die quickly.

Figure 7.10(a)-(b) shows the energy (J) consumption of the network per round for 100 and 400 sensor nodes respectively. It can be observed that DEECR performs better than LEACH, DEBR and CBCRA in terms of energy consumption of the sensor nodes. This is because of the proper next hop relay node selection to reduce the energy consumption in data routing phase.

The performance of the algorithms is also tested in terms of total data packets sent to the BS until the last sensor node becomes inactive or disconnected. It is observed that DEECR sends more data packets to the BS than LEACH, CBCRA and DEBR. This is due to the fact that the algorithm provides guaranteed connectivity between the CHs, where as CBCRA forms clusters without assuring the connectivity. As a result a few CHs are not able to send the aggregated data packets to the base station. On the other hand, DEBR performs poor in total packet transfer. The rationale is that in order to balance the energy consumption, some sensor nodes in DEBR tries to send the data packets to the base station through other sensor nodes which may be

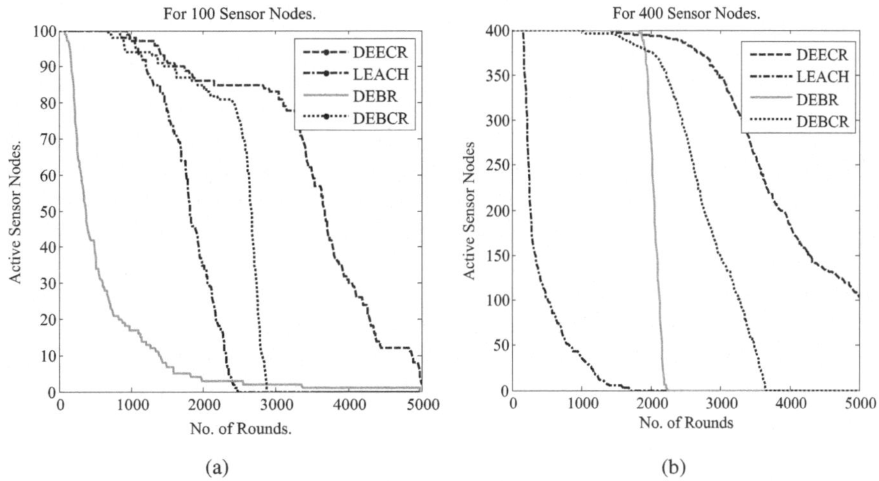

FIGURE 7.9: Comparison in terms of active sensor nodes for (a) 100 and (b) 400 sensor nodes

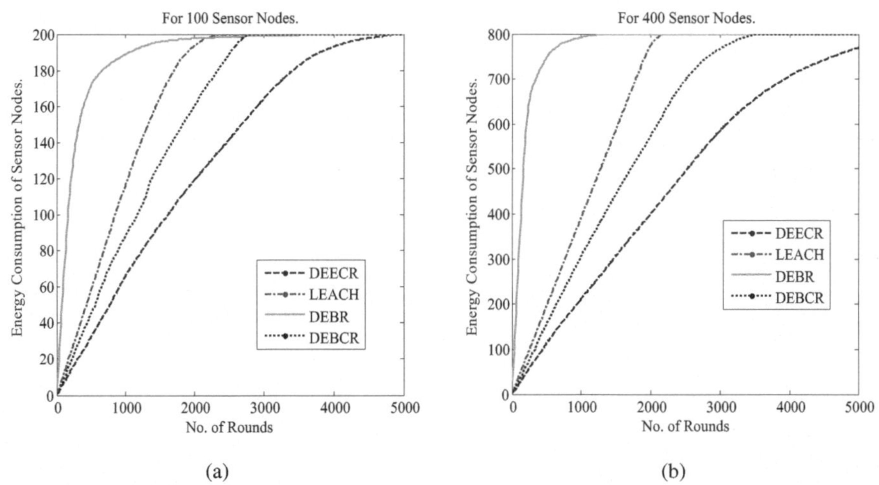

FIGURE 7.10: Comparison in terms of energy consumption for (a) 100 and (b) 400 sensor nodes

in opposite direction of the base station and in the same way the packets finally may reach to a particular region from where there may not be any forward path reaching to the base station. As a result, a massive amount of energy consumes without data packets reaching to the final destination, i.e., the BS.

7.5 Chapter summary

In this chapter, first two algorithms are presented for distributed clustering and routing in WSNs. In the first algorithm, cluster heads (CHs) are selected from normal sensor nodes on the basis of composite weight value of their residual energy and neighborhood cardinality. All non-CH sensor nodes select a CH within their communication range, considering a cost value of the CHs. All sensor nodes use single-hop communication to communicate with its CH and all CHs use multi-hop communication to send the aggregated data to base station. In routing, the cost of each path towards base station to select a proper route which can balance the energy of the sensor nodes is also measured. However, the algorithm does not assured the connectivity between selected CHs. Next another distributed algorithm called DEECR is presented. DEECR ensures connectivity between the CHs during data routing phase. Proper cluster range is derived to ensure the connectivity. It is also shown that the routing algorithm consumes comparably lesser energy in forwarding the data packets. It is important to note that the algorithms can also work for the dynamically changed WSNs with arbitrary addition and deletion of the sensor nodes. This is due to the fact that the algorithm is distributed in nature and executed in each round using local information and network structure.

However, mobility of the nodes is not considered in all of the above described algorithms. Moreover, for fault tolerance of the algorithms, only permanent failure of the sensor nodes is considered. Therefore, research can be carried out to design energy aware distributed clustering and routing for mobile sensor network with partial and transient failure of the sensor nodes.

Chapter 8

Fault Tolerant Algorithms

Chapter 6

Chapter 8

Fault Tolerant Algorithms

8.1 Introduction

In many applications, sensor nodes are deployed in hostile or harsh environments. Therefore, these sensor nodes are susceptible to failure due to several factors, including complete depletion of battery power, malfunctioning of hardware components (such as processing unit, transceiver, etc.), or destruction by an external event. The failure of sensor nodes makes some parts of the target region uncovered and thus creates network partitioning. Therefore, it affects the overall network lifetime and degrades the performance of the network. However, the failure of single or some sensor nodes may not affect the overall performance of the network as the WSN is intended to gather information from a group of sensor nodes rather than a single or some sensor nodes. Nevertheless, the failure of a CH in cluster based WSNs has serious impact on the network performance. This is because the failure of CH not only disrupts the communication with member sensor nodes, but also with other CHs as they are involved in routing aggregated data to the sink through other CHs. Therefore, fault tolerance of WSNs with respect to CH failure is an extremely important issue in the development of a large scale WSN. A possible solution to this problem is to replace the faulty nodes, but it leads to poor utilization of the resources. Moreover, replacement of the nodes may not be possible in many applications. Therefore, in order to continue functionality of the WSNs, the clustering and routing algorithms should cope with fault tolerant aspect, especially when some CHs fail to work [112, 115].

In this chapter, first a distributed fault-tolerant clustering and routing (DFCR) algorithm is presented. The algorithm emphasizes on the failure of the CHs. The DFCR addresses the issues of energy efficiency. In the clustering phase, sensor nodes select their CH based on a cost function that consists of residual energy of the CH. This is also based on the distance between sensor node to the CH and the distance from the CH to the base station. The algorithm takes care of the sensor nodes which have no CH within their communication range. The DFCR presents a distributed run time recovery of the faulty cluster members due to sudden failure of the CH. To achieve fault tolerance, the method adopts neither redundant deployment of the CHs nor reclustering approach. In data routing phase, the CHs select their next hop neighbor CH in such a way that their energy consumption will be balanced and minimized. The algorithm can tolerate the sudden failure of CHs in the routing path.

Finally, a distributed algorithm called EEBR (Energy Efficient and Balanced Routing) is also presented. It is also shown that the algorithm is fault tolerant. The sensor nodes select their next hop neighbor sensor node in such a way that their energy consumption is balanced as well as minimized. The algorithm can tolerate the sudden failure of the sensor nodes in routing path.

8.2 Fault tolerant algorithms

8.2.1 Clustering algorithm

For better understanding of the algorithms, some terminologies are described as follows:

1. $ComCH(s_i)$: The set of all those gateways, that are within the communication range (R_S) of sensor node s_i. In other words,

$$ComCH(s_i) = \{g_j | dis(s_i, g_j) \leq R_S \wedge g_i \in \xi\} \tag{8.1}$$

   Therefore, s_i can be assigned to any one of the gateways from $ComCH(s_i)$, where $ComCH(s_i) \subseteq \xi$.

2. $Neighbor(s_i)$ is the set of all those sensor nodes, that are within the communication range of node s_i. Therefore,

$$Neighbor(s_i) = \{s_j | dis(s_i, s_j) \leq R_S \wedge s_j \in \{S - s_i\}\} \tag{8.2}$$

3. $Com(g_i)$: The set of gateways that are within communication range of g_i. The BS may also be a member of $Com(g_i)$. In other words,

$$Com(g_i) = \{g_j | \forall g_j \in \{\xi + g_{m+1}\} \wedge dis(g_i, g_j) \leq d_{max}\} \tag{8.3}$$

4. $HCount(g_i)$ denotes the number of next hops required to reach to the BS from g_i. If g_i can directly communicate with BS, then $HCount(g_i)$ is one. Therefore, $HCount(g_i)$ can be recursively defined as

$$HCount(g_i) = \begin{cases} 1, & \text{if next hop is } g_{m+1} \text{ (i.e.,BS)} \\ 1 + HCount(g_i), & \text{next hop is } g_j \text{ (i.e., other than BS)} \end{cases} \tag{8.4}$$

5. $BackCH(g_i)$: The set of gateways from $Com(g_i)$ with a lesser hop count than g_i. Therefore,

$$BackCH(g_i) = \{g_k | \forall g_k \in Com(g_i) \wedge Hcount(g_i) < Hcount(g_k)\} \tag{8.5}$$

Depending on the communication range and connectivity between the sensor nodes and the gateways, we have defined few kinds of sensor nodes in the system as follows [170].

Definition 1 (covered node and covered set). *Covered nodes are those sensor nodes, which have at least one gateway within its communication range. Covered set is the set of all covered nodes in the WSN. We refer this set as CO_{set}. It is obvious to note that a sensor node s_i belongs to CO_{set}, if it satisfies the following criteria.*

$$s_i \in CO_{set} \Leftrightarrow [\{\exists g_j | g_j \in ComCH(s_i) \wedge g_j \in \xi\}] \tag{8.6}$$

Definition 2 (uncovered node and uncovered set). *Uncovered nodes are those sensor nodes, which have no gateway within its communication range. Uncovered set is the collection of all uncovered nodes in the WSN. We refer this set as $UnCO_{set}$. A sensor node s_i belongs to $UnCO_{set}$, if it satisfies the following criteria.*

$$s_i \in UnCO_{set} \Leftrightarrow [s_i \notin CO_{set}] \tag{8.7}$$

Definition 3 (backup nodes and backup set). *Backup nodes of an uncovered sensor node s_i are all the covered sensor nodes which are within communication range of s_i. Backup set of an uncovered sensor node s_i is the set of all backup nodes of s_i. We refer this set as $BackupSet(s_i)$. Therefore,*

$$BackupSet(s_i) = \{s_j | s_j \in Neighbor(s_i) \wedge s_j \in CO_{set}\} \tag{8.8}$$

Definition 4 (alive and inactive sensor node). *Alive sensor nodes are those sensor nodes, which have some residual energy and can send the sensed data to the CH directly or indirectly. Sometimes, few sensor nodes may have residual energy but unable to communicate with its CH due to CH failure. These sensor nodes are called inactive sensor nodes.*

8.2.1.1 Bootstrapping

At the beginning, all the nodes undergo bootstrapping process, in which the BS broadcasts a HELLO message at a certain power level. In this way, each gateway can compute the approximate distance to the BS based on the received signal strength indicator [104, 171]. It helps the gateways to select the proper power level to communicate with the BS. Then the BS floods a *HopPacket* which consists of a counter indicating the hop value. Initially the counter is set to one. After receiving the packet, a gateway stores the counter value, increments it by one and transmits it to their neighbor gateways. This transmission will continue until all the CHs are aware of their hop values for their use in data routing phase. Note that as the communication is through the wireless link, the gateways which have already transmitted their data packets, will also receive packets from their neighbor CHs. Therefore, after receiving the packet a gateway will check the counter value with its own stored one. If the received counter value is less than the stored value, then only the gateway increments the counter value, restores and retransmits it; otherwise it will ignore the received packets. This will help to avoid the redundant transmission and loop creation.

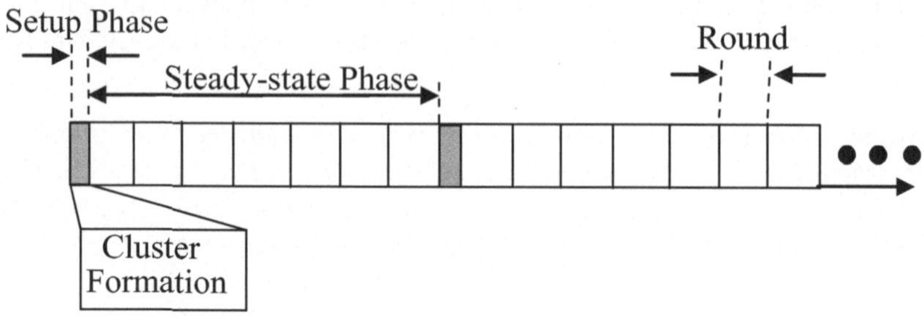

FIGURE 8.1: Network setup

Network setup consists of a setup phase followed by the steady-state phase as shown in Figure 8.1 and illustrated as follows. In the setup phase, clusters are formed using the algorithm discussed later (Algorithm 10). Steady-state phase divides the schedule of the network into multiple rounds of fixed duration. In each round, the gateways receive the sensed data from the cluster members and aggregate them to transfer it to the base station using the distributed multi-hop routing algorithm discussed later. The steady-state phase is composed of some pre-specified rounds, say 50 or 75 rounds. Setup and steady-state phases are repeated until all the sensor nodes become inactive. Now, the details of cluster formation and inter cluster routing is presented in the following sections subsequently.

8.2.1.2 Distributed clustering algorithm

In the setup phase, gateways broadcast a HELLO message within the communication range of a sensor node. The HELLO message consists of gateway ID, residual energy and distance to the base station. After a certain timeout, if a sensor node s_i has received at least one HELLO message, then it becomes a member of CO_{set}, otherwise it becomes a member of $UnCO_{set}$. Each sensor node $s_i \in UnCO_{set}$ broadcasts a HELP message for backup. All the sensor nodes of $Neighbor(s_i)$ which are belonged to CO_{set} send a reply to form $BackupSet(s_i)$. If the backup set is not empty i.e., $|BackupSet(s_i)| \neq 0$, then s_i uses s_j from $BackupSet(s_i)$ as a relay with highest residual energy through which s_i could send the data to the CH. In other words, s_i uses s_j as a relay if

$$E_{Residual}(s_j) = max\{E_{Residual}(s_k) | \forall s_k \in Backset(s_i)\} \tag{8.9}$$

Thus the sensor nodes, which are not covered due to random deployment or sudden failure of CHs, are assigned to a CH through multi-hop communication. Now, the sensor nodes $s_j, \forall s_j \in CO_{set}$, join a CH based on the cost value. Let $CHCost(g_i, s_j)$ be the cost of the cluster head g_i for s_j. To define the cost function, we consider the following parameters.

1. *Residual energy of CH:* A sensor node should join that CH which has higher residual energy than any other CH within its cluster range. Therefore,

$$CHCost(g_i, s_j) \propto E_{Residual}(g_i) \tag{8.10}$$

2. *Distance from sensor node to CH:* As a non-CH sensor nodes consume maximum energy to communicate with its CH, sensor node should join the nearest CH. The shorter the distance higher is the chance to join. Therefore,

$$CHCost(g_i, s_j) \propto \frac{1}{dis(s_j, g_i)} \tag{8.11}$$

3. *Distance from CH to base station:* Gateways are capable of long-haul communication compared to sensor nodes and can communicate directly with the BS. Thus, the gateways that are far away from the BS, consume more energy for long-haul communication to the BS. Therefore, the number of cluster members of these CHs should be less than that of the CHs which are nearer to the BS. In other words,

$$CHCost(g_i, s_j) \propto \frac{1}{dis(g_i, g_{m+1})} \tag{8.12}$$

By combining equations (8.10), (8.11) and (8.12), we obtain

$$CHCost(g_i, s_j) \propto \frac{E_{Residual}(g_i)}{dis(s_j, g_i) \times dis(g_i, g_{m+1})} \tag{8.13}$$

$$i.e., CHCost(g_i, s_j) = K \times \frac{E_{Residual}(g_i)}{dis(s_j, g_i) \times dis(g_i, g_{m+1})} \tag{8.14}$$

Where, K is a proportionality constant. It is noteworthy that we take the cost value of the CHs only for comparison purpose. So, the value of K does not hamper our objective and without loss of generality, we can assume $K = 1$. Therefore,

$$CHCost(g_i, s_j) = \frac{E_{Residual}(g_i)}{dis(s_j, g_i) \times dis(g_i, g_{m+1})} \tag{8.15}$$

Now s_j selects the cluster head say g_i with highest cost value. Therefore,

$$CHCost(g_i, s_j) = Max\{CHCost(g_k, s_j) | \forall g_k \in ComCH(s_j)\} \tag{8.16}$$

The sensor nodes ($s_j, \forall s_j \in CO_{set}$) select a CH to join using equation (8.16) and then sends a JOIN_REQ message to its selected CH using a non-persistent CSMA/CA MAC protocol. After completion of cluster formation, each CH sets up a TDMA schedule for its member nodes and transmits this schedule to the member sensor nodes. This TDMA schedule ensures that no collisions will occur during communication between sensor nodes and CH. It also ensures that the radio component of each sensor node will be turned off except their transmit time. Thus, it reduces the energy consumption. The cluster formation algorithm for sensor node s_i is given in algorithm 10. This algorithm simultaneously runs by each sensor node $s_i, \forall s_i, 1 \leq i \leq n$.

Algorithm 10 Cluster-Formation

1: $ComCH(s_i) = \{\}$
2: $BackupSet(s_i) = \{\}$
3: **while** (*not timeout*) **do**
4:     **if** (s_i receives a HELLO message from g_k) **then**
5:         s_i becomes a member of CO_{set}
6:         $ComCH(s_i) = ComCH(s_i) \cup g_k$
7:     **end if**
8: **end while**
9: **if** (no HELLO message is received by s_i) **then**
10:     s_i becomes a member of $UnCO_{set}$
11: **end if**
12: **if** (($s_i \in UnCO_{set}$)) **then**
13:     s_i broadcasts a HELP message for backup.
14:     **while** (s_i receiving reply from s_j) **do**
15:         $BackupSet(s_i) = BackupSet(s_i) \cup s_j$
16:     **end while**
17:     **if** ($|BackupSet(s_i)| \neq 0$) **then**
18:         s_i uses s_j from $BackupSet(s_i)$ as a relay with highest residual energy to send the data to the CH.
19:     **end if**
20: **else**
21:     s_i calculates cost of all these CHs according to equation (8.15) and joins the CH with the highest cost value.
22:     **if** (s_i receives any HELP message) **then**
23:         Send reply message.
24:     **end if**
25: **end if**

Lemma 1. *The Cluster-Formation algorithm has worst case time complexity of $O(n)$ where n is the number of sensor nodes in the network.*

Proof. Line 1-2 and line 9-11 of the algorithm takes constant time. Line 3-8 can be executed in $O(m)$ time as a sensor node may have m number of gateways within its communication range in the worst case. In the line 12-25, sensor node resets its backup set (line 14-16) in $O(n)$ time and selects a CH in $O(m)$ processing time (line 21). The overall execution time of line 12-25 is $O(m)+O(n)$, i.e., $O(n)$ in a worst case situation as $n > m$. Therefore, the overall processing time of the cluster formation algorithm is $O(n)$ in the worst case as line 12-25 dominates all other steps of the algorithm. □

Lemma 2. *The Cluster-Formation algorithm has worst case message exchange complexity of $O(1)$ per sensor node, i.e., $O(n)$ for the whole network having n sensor nodes.*

Algorithm 11 Clustering-Fault-Tolerance

1: s_i broadcast a HELP message within its communication range.
2: $ComCH(s_i) = \{\}$;
3: $BackupSet(s_i) = \{\}$;
4: **while** (s_i receiving reply) **do**
5:     **if** (reply is from gateway g_k) **then**
6:         s_i becomes a member of CO_{set}
7:         $ComCH(s_i) = ComCH(s_i) \cup g_k$
8:     **else**
9:         **if** (reply is from sensor node s_j) **then**
10:             $BackupSet(s_i) = BackupSet(s_i) \cup g_j$
11:         **end if**
12:     **end if**
13: **end while**
14: **if** ($s_i \in UnCO_{set}$ & $|BackupSet(s_i)| \neq 0$) **then**
15:     s_i uses s_j from $BackupSet(s_i)$ as a relay with highest residual energy to send the data to the CH.
16: **else**
17:     s_i calculates cost of all CHs from $ComCH(s_i)$ according to equation (8.15) and joins the CH with the highest cost value.
18: **end if**

Proof. In cluster formation phase, a sensor node calculates the cost value of the CHs within its communication range and sends a join request to the selected CH. However, the sensor nodes which have no CHs within their communication range, broadcast a HELP request message (line 13). If it finds any sensor node with helping hand, then it sends a join request message to join that cluster via the helping sensor node with multi-hop communication. Therefore, sensor node has to send only two messages in the worst case for the formation of the cluster. Hence, message complexity of each sensor node is constant, i.e., $O(1)$. Thus, overall message exchange complexity for the whole network is $O(n)$. □

8.2.1.3 Fault tolerance

During the steady state phase, at any time a CH may fail due to depletion of energy or some damages. The fault can be detected whenever the member sensor nodes do not receive any data acknowledgment receipt from the CH and also the failure of the CH can be confirmed from the neighbor sensor nodes of the same cluster [172]. After detection of the fault, the member sensor nodes of the faulty cluster broadcast a HELP message within its communication range. The sensor nodes $s_j, \forall s_j \in Neighbor(s_i)$, from other cluster and the CHs $g_k, \forall g_k \in ComCH(s_i)$, reply this HELP message. If the a sensor node receives any reply from other CHs then it remains as elements of CO_{set}, otherwise it becomes the elements of $UnCO_{set}$. The sensor nodes $s_i, \forall s_i \in UnCO_{set}$, those have received reply from other sensor nodes

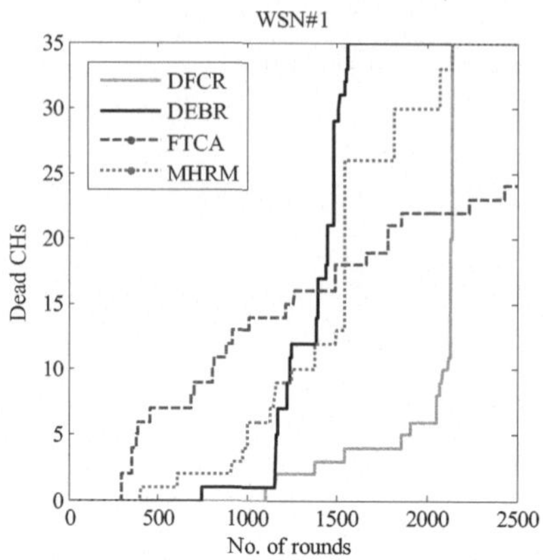

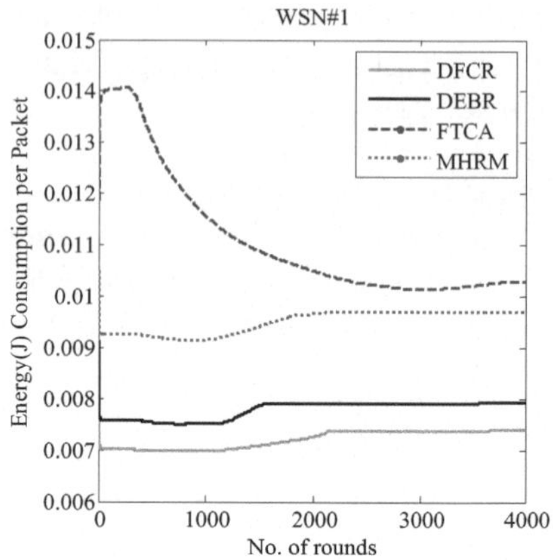

FIGURE 8.2: Comparison in terms of (a) number of dead CHs and (b) average energy consumption per packet

reset their $BackupSet(s_i)$. Now, $s_j, \forall s_j \in CO_{set}$, joins a CH considering the cost value as in equation (8.15) and $s_i, \forall s_i \in UnCO_{set}$, uses $s_j, s_j \in BackupSet(s_i)$, as a relay with highest residual energy to send the data to the CH. The fault tolerance algorithm for sensor node s_i is given in Algorithm 11. This algorithm is individually run by the sensor nodes from the faulty cluster.

The algorithm is compared with other fault tolerant clustering algorithm (FTCA) [39], the distributed algorithm DEBR [104] and MHRM [163]. Note that DEBR and MHRM are basically routing algorithms without any clustering phase. Therefore, both these algorithms are run along with the described cluster formation algorithm for the fair comparison.

Figure 8.2(a) shows the comparison results in terms of number of dead CHs round by round. Note that the number of dead CHs includes all the CHs which are not operational due to complete energy depletion or device failure following Weibull reliability function. It can be observed that DFCR performs better than the FTCA, DEBR and MHRM in terms of the number of dead CH during the network lifetime. The rationale behind the poor performance of FTCA is that, the CHs use direct communication with BS which leads to the initial death of the CHs due to huge energy consumption for long haul transmission. On the other hand, in order to balance the energy consumption some CHs in DEBR try to send the data packets to the base station through other CHs which may be in the opposite direction of the base station and in the same way the packets finally may reach to a particular region from where there may not be any forward path reaching to the base station. As a result, a massive amount of energy is consumed without data packets reaching to the final destination, i.e., the BS. To minimize the hop count, the CHs in MHRM select longest distance CH towards the BS. As a result, the CHs die quickly for huge energy consumption due to long haul communication. Whereas, DFCR prevents the initial death of the

CHs through properly balance the energy consumption and energy efficient selection of next hop CHs. Moreover, if a few CHs die due to complete energy depletion or device failure, DFCR finds a route to the BS until there exists a route and the network becomes connected for a longer period of time. Note that the long life of the CHs helps to reach more number of data packets to the BS. It is also observed that the DFCR performs better than the existing algorithms in terms of the number of packets received by the BS until the last gateway becomes inactive or disconnected. Figure 8.2(b) shows energy efficiency. As the number of live CHs in DFCR is higher than the others, it transmits a higher number of data packets to the BS. The average energy consumption per packet in DFCR is comparably lower than the others as shown in Figure 8.2(b).

The algorithms are also tested in terms of number of inactive sensor nodes in the network. A sensor node is considered as active if its existing energy is not zero and also there must be at least one gateway within its communication range. Sometimes few gateways die quickly for improper energy balancing. As a result, few sensor nodes are unable to find any gateway within their range, though it still may have some existing energy. In this scenario, this type of sensor node is also considered as inactive. It is observed that the number of inactive sensor nodes round by round in DFCR is increasing lesser than the existing algorithms. This is due to the fact that DFCR takes care about the energy consumption of the normal sensor nodes.

8.2.2 Routing algorithm

8.2.2.1 Bootstrapping

At the beginning, all the nodes undergo bootstrapping process, in which the BS floods a *Hop-Packet* which consists of a counter indicating the hop value. Initially the counter is set to one. After receiving the packet, a sensor node stores the counter value, increments it by one and transmits it to their neighbor sensor nodes. This transmission will continue until all the sensor nodes are aware of their hop values for their use in data routing phase. Note that as the communication is through the wireless link, the sensor node which have already transmitted their packet, will also receive packets from its neighbor sensor nodes. Therefore, after receiving the packet a sensor node will check the counter value with its own stored one. If the received counter value is less than the stored value, then only the sensor node increments the counter value, restores and retransmits it; otherwise it will ignore the received packets. This will help to avoid the redundant transmission and loop creation. After the fixed time out, the nodes set their hop value as final. Now the details of routing algorithm is presented in the following section.

8.2.2.2 Distributed routing

Here, sensor nodes select the next-hop relay nodes in such a way that it requires less energy and also takes care of balancing the energy consumption. This is implemented as follows. The sensor nodes are aware of their hop count. As the communication is through a wireless link, sensor nodes are also aware of some information about their neighbor sensor nodes such as the distance from their neighbors, neighbor remaining energy and hop count as in [104, 105]. Generally a sensor node may have many possible next-hop relay nodes within its communication range. However, in our approach only those neighbor sensor nodes are eligible to serve as a relay node which has lesser hop count. Now, a sensor node s_j chooses the next hop relay node s_r such that $H_C(s_i) > H_C(s_r)$ based on the cost function. We denote the cost function as $Cost(s_j, s_r)$ which indicates the cost of selecting the s_r as next hop relay node by the s_j. To define the cost function, we consider the following parameters [173].

1. ***Energy density***: Our objective is to balance the energy consumption of the network. The general principle behind energy balancing is that the data packets should move forward to the base station through the region of the network where sensor nodes have higher remaining energy. In order to achieve this objective we calculate energy density of a sensor node and the data packets are force to flow through the higher density region. Energy density of a sensor node s_k can be calculated as follow.

$$\begin{aligned} &E_{Density}(s_k) \\ &= \sum \{E_{Residual}(s_i), \forall s_i \in Neighbor(s_k) \\ &\wedge H_C(s_i) < H_C(s_k)\} + E_{Residual}(s_k) \end{aligned} \tag{8.17}$$

   The higher the $E_{Density}(s_k)$, the higher is the chance of getting selected as a relay node. Therefore, the cost function is proportional to the $E_{Density}(s_k)$. In other words,

$$Cost(s_j, s_k) \propto E_{Density}(s_k) \tag{8.18}$$

2. ***Residual energy***: Although s_k may have sufficient energy density due to comparably higher residual energy of its next-hop neighbors, it may have lesser remaining energy to serve as a relay node. In order to balance the energy consumption of the network, s_k should have also sufficient residual energy. Higher the residual energy, the higher is the chance of getting selected as a relay node. Therefore, the cost function is proportional to the $E_{Residual}(s_k)$. In other words,

$$Cost(s_j, s_k) \propto E_{Residual}(s_k) \tag{8.19}$$

3. ***Distance to receiver***: As s_j consumes its energy due to long-haul transmission with s_k, the distance between s_j and s_k should be lesser to reduce energy consumption of s_j. The lesser the transmission distance, the higher is the chance of getting selected as a relay node. Therefore, the cost function is inversely proportional to the $dis(s_j, s_k)$. In other words,

$$Cost(s_j, s_k) \propto \frac{1}{dis(s_j, s_k)} \tag{8.20}$$

By combining equations (8.18), (8.19) and (8.20), we obtain

$$Cost(s_j, s_k) \propto \frac{E_{Density}(s_k) \times E_{Residual}(s_k)}{dis(s_j, s_k)}$$

$$Cost(s_j, s_k) = K \times \frac{E_{Density}(s_k) \times E_{Residual}(s_k)}{dis(s_j, s_k)}$$

where K is a proportionality constant. It is noteworthy that, we take the weight value of the sensor nodes only for comparison purpose. Therefore, the value of K does not hamper our objective and without loss of generality, we can assume $K = 1$. Therefore,

$$Cost(s_j, s_k) = \frac{E_{Density}(s_k) \times E_{Residual}(s_k)}{dis(s_j, s_k)} \tag{8.21}$$

Note that, in the above cost function (equation (8.21)), the three used parameters have different range of their value. Therefore, in order to give equal importance to the parameters, their values are normalized within the range of (0,1). Now, s_j selects the s_k having maximum cost value as relay node. Therefore,

$$Cost(s_j, s_k) \quad = \quad \text{Max}\ \{Cost(s_j, s_r) | \forall s_r \in Neighbor(s_j) \wedge H_C(s_j) > H_C(s_r)\} \tag{8.22}$$

The sensor nodes which are within communication range (d_{max}) of the BS will not execute the data routing algorithm. They sends their data directly to the base station. Other sensor nodes forward the data to the next-hop relay node using a non-persistent CSMA/CA MAC protocol. The next hop relay node selection algorithm for s_i is given in Algorithm 12.

Remark 1. *The above cost function efficiently addresses both the issues of energy balancing as well as energy consumption of the network. The first and second parameters ($E_{Density}(s_k)$ and $E_{Residual}(s_k)$) of the cost function, take care of the energy balancing of the network. The third parameter ($dis(s_j, s_k)$) tries to reduce the energy consumption of the network. Thereby, it helps the routing algorithm to be an energy efficient.*

Lemma 3. *The EEBR-Next-hop-Selection algorithm has worst case time complexity of $O(n)$ without any fault and $O(n^2)$ with occurring of faults where n is the number of sensor nodes.*

Proof. Line 2 through Line 6, require constant time. To select a next hop relay node, a sensor node has to compare the cost value of its neighbor sensor nodes in Line 7 which takes $O(n)$ time, assuming that a sensor node may have $n - 1$ neighbor sensor nodes in worst situation. Therefore, next hop selection takes $O(n)$ time in the worst case. However, whenever fault occurs then a sensor node increases its hop count by 1 and Line 4 through Line 7 are repeated requiring $O(n^2)$ time as a sensor node can increase its hop count a maximum of $n - 1$ times. □

Lemma 4. *The EEBR-Next-hop-Selection algorithm is loop-free in routing the data to sink.*

Algorithm 12 EEBR-Next-Hop-Selection

Input: (1) Hop count of s_i.
(2) $\forall s_k \in Neighbor(s_i)$: energy density, residual energy, hop count, distance to s_k.
Output: Next-hop relay node (*Relay*) of s_i.

1: **if** $(dis(s_i, s_{N+1}) \leq dmax)$ **then**
2: $\quad$ $Relay = s_{N+1}$ /* s_{N+1} indicates BS.*/
3: **else**
4: $\quad$ $N_i = Neighbor(s_i)$
5: $\quad$ $max = -1.0$
6: $\quad$ $Relay = -1$
7: $\quad$ **while** $(N_i \neq NULL)$ **do**
8: $\quad\quad$ Select a s_r from N_i.
9: $\quad\quad$ **if** $(Cost(s_i, s_r) > max$ && $H_C(s_i) > H_C(s_r))$ **then**
10: $\quad\quad\quad$ $Relay = s_r$
11: $\quad\quad\quad$ $max = Cost(s_i, s_r)$
12: $\quad\quad$ **end if**
13: $\quad\quad$ $N_i = N_i - s_r$
14: $\quad$ **end while**
15: $\quad$ **if** $(Relay == -1$ && $H_C(s_i) \leq N)$ **then**
16: $\quad\quad$ $H_C(s_i) = H_C(s_i) + 1$
17: $\quad\quad$ Repeat Line 4 to Line 7.
18: $\quad$ **end if**
19: **end if**
20: Stop.

Proof. Let us assume that there exists a loop $s_a \rightarrow s_b \rightarrow s_c \cdots s_r \rightarrow s_a$ in the routing graph. As per the algorithm, a node s_i selects $s_k (i \neq k)$ as a next hop relay node only when it fulfills the condition $H_C(s_i) > H_C(s_k)$. Therefore, if we consider the graph up to the point when loop is not still created then we find that $H_C(s_a) > H_C(s_b) > H_C(s_c) > \cdots > H_C(s_r)$. As per the assumption, a loop is created because s_r selects s_a as its next hop relay node, which implies that $H_C(s_r) > H_C(s_a)$. But this contradicts $H_C(s_a) > H_C(s_b) > H_C(s_c) > \cdots > H_C(s_r)$. □

Fault tolerance

Note that the above next-hop selection algorithm has the ability to cope up with the failure of the sensor nodes as follows. Whenever a sensor node does not receive any acknowledgment from its neighbor sensor node, it detects the neighbor as failed node. This can also be confirmed by asking the other neighbor nodes. Note that failure of one or more sensor nodes does not create any problem in data routing phase if there is still a live sensor node with lesser hop count to serve as a relay node. But, there may be a situation when a sensor node may not have any next-hop sensor node with lower hop count. Then the sensor node increases its hop count by one and broadcasts the same to its neighbors to update their hop count. Now the sensor node searches for other sensor nodes having lesser hop count than itself. If it finds some

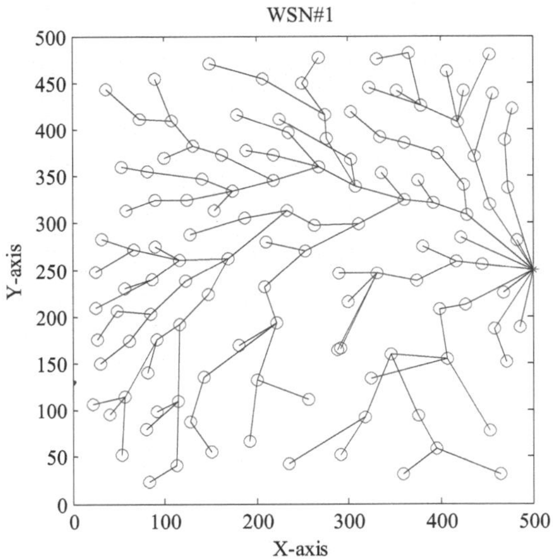

FIGURE 8.3: Routing paths from the sensor nodes in 1st round for 150 sensor nodes in WSN#1

nodes with the lesser hop count, then it selects one of them as a next hop relay node considering the cost value as described in equation (8.21).

The simulation run was carried out with various network scenarios, namely WSN#1 and WSN#2. WSN#1 has the sensing field of 500×500 square meters area and the position of the base station is taken in the side of the region with coordinate (500,250). Here, simulation is carried out for 150 sensor nodes. Figure 8.3 shows the paths for data routing from each sensor nodes to the BS in the 1st round of the network. It can be observed from Figure 8.3 that the tendency of the data flow is towards densely deployed zone of the network. This is due to the next-hop relay node selection strategy of the algorithm. In order to balance the energy consumption of the network, the algorithm forces the data to flow towards the densely energy region of the network.

The second scenario (i.e., WSN#2) has the same sensing field and same position of the base station. But, here a real life complex network scenario is assumed where a few regions of the sensing field are created as holes, i.e., which are not covered by the sensor nodes. The simulation run is performed for 100 sensor nodes and Figure 8.4 shows the routing paths for each sensor node to the BS in the 1st round of the network.

First, the algorithms are executed to compare their energy efficiency for 200 sensor nodes in both the network scenarios, WSN#1 and WSN#2. An algorithm is said to be energy efficient if it consumes lesser energy to reach a data packet to the BS. Figure 8.5(a)-(b) shows the comparison of the algorithms in terms of average energy consumption per data packets to reach to the BS in WSN#1 and WSN#2 respec-

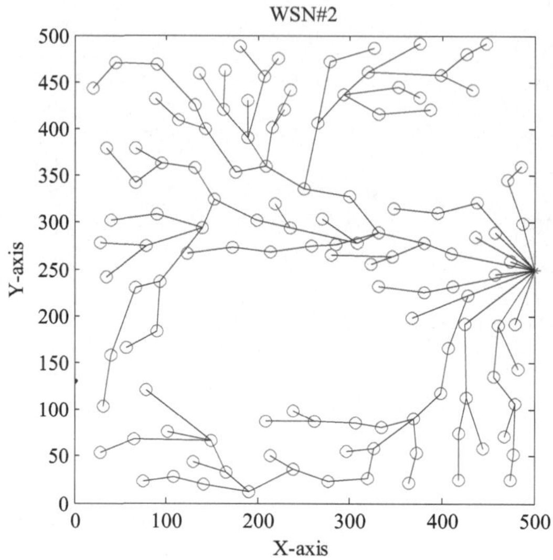

FIGURE 8.4: Routing paths from the sensor nodes in 1st round for 150 sensor nodes in WSN#2

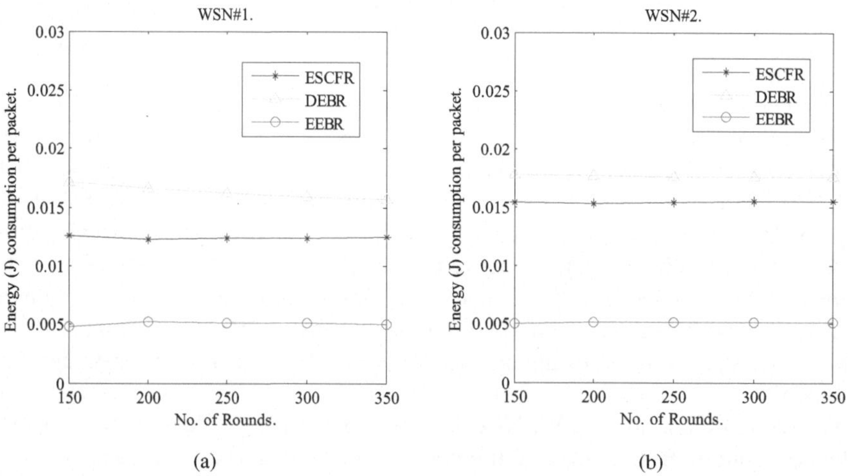

FIGURE 8.5: Comparison of average energy (J) consumption per packet in (a) WSN#1 and (b) WSN#2

tively. It can be observed that the algorithm EEBR takes significantly lesser energy than DEBR and ESCFR to sends a data packet to the BS.

Note that, both the algorithms DEBR and ESCFR consider that all the neighbors are within the fixed transmission range. Therefore, a sensor node uses the fixed higher

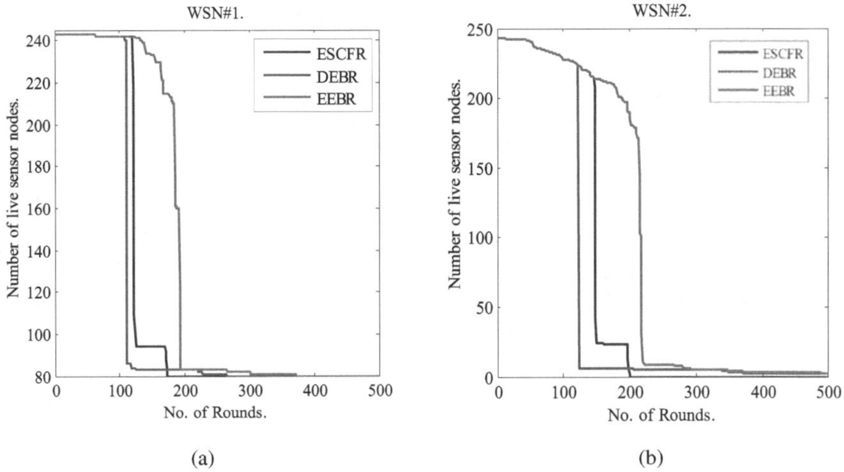

FIGURE 8.6: Comparison in terms of number of live sensor nodes in (a) WSN#1 and (b) WSN#2

transmission level to communicate with its neighbor sensor node, even if they are very close to each other. Furthermore, in order to balance the energy consumption, some sensor nodes try to send the data packets to the base station through other sensor nodes which may be in opposite direction of the base station and in the same way the packets finally may reach to a particular region from where there may not be any forward path reaching to the base station. As a result, a massive amount of energy consumes without data packets reaching to the final destination, i.e., the BS.

Next, the algorithms are compared in terms of the number of live sensor nodes in the network. A sensor node is considered to be alive if it has sufficient residual energy and also connected with the network. Note that, here the sudden failure of sensor nodes due to device failure is considered to follow Weibull reliability function. Figure 8.6(a)-(b) show the comparison of the algorithms in terms of number of live sensor nodes in WSN#1 and WSN#2 respectively. The experiments are performed for 240-250 sensor nodes for both of the network scenarios. It can be observed from Figure 8.6 that, EEBR performs better than the others in terms of number of live sensor nodes in the network. The algorithms are also tested in terms of total number of data packets received by the BS. It is observed that the EEBR sends significantly higher number of data packets to the BS.

8.3 Chapter summary

In this chapter, first a distributed fault-tolerant clustering and routing algorithms jointly called DFCR for wireless sensor networks (WSNs) is presented. An efficient mechanism for runtime recovery of the faulty cluster members in a distributed way is also presented. The routing algorithm has been shown to cope with sudden failure of the CHs in the routing path without creating any loop. It has been shown that the algorithm is more efficient in terms of number of dead CHs, energy consumption, number of inactive sensor nodes and so on.

Then, a distributed routing algorithm called EEBR for wireless sensor networks (WSNs) is presented. The algorithm has been shown to build a trade-off between energy-efficiency and energy-balancing of the sensor nodes. The EEBR uses energy density of the network to balance the energy consumption. It is shown that the EEBR can cope with sudden failure of the sensor nodes in the routing path without creating any loop. The simulation results in diverse network scenarios have shown that the algorithm is more efficient than the existing algorithms in terms of energy consumption as well as energy balancing.

However, mobility of the nodes is not considered in all of the above described algorithms. Moreover, for fault tolerance of the algorithms, only permanent failure of the sensor nodes is considered. Therefore, research can be carried out to design energy aware distributed clustering and routing for mobile sensor network with partial and transient failure of the sensor nodes.

Chapter 9

Unequal Clustering and Mobile Sink-Based Routing

Chapter 9

Unequal Clustering and Mobile Sink-Based Routing Algorithms

9.1 Introduction

In multi-hop communication, the sensor nodes that are closer to the BS inevitably perform data forwarding. As a result, the energy consumption of these sensor nodes is comparably higher than that of the other sensor nodes. This imbalanced energy consumption leads to the premature death of a certain number of nodes near to the BS, and causes the hot spot problem. One such situation is shown in Figure 9.1. There are many solutions to resolve this problem. Here, we present two methods, i.e., unequal clustering and use of a mobile sink.

9.2 Unequal clustering algorithms

9.2.1 Energy-efficient unequal clustering (EEUC)

EEUC [166] is a distributed unequal clustering algorithm, where cluster heads are selected based on the residual energy of the sensor nodes. In order to distribute the energy consumption of the network, the overhead of being a CH is rotated among the sensor nodes in each data gathering round same as it is done in LEACH.

At first, several cluster heads are selected tentatively to compete for final CH selection. Every node tentatively becomes a cluster head with some probability T, which is a predefined threshold. The sensor nodes that are the tentative CHs have a competition range R_{comp}, which is a function of their distance to the base station. Let R_0^{comp} be the maximum predefined competition radius. Then the competition range R_{comp} of s_i is calculated as follows:

$$R_{comp}(s_i) = (1 - c \times \frac{d_{max} - d(s_i, BS)}{d_{max} - d_{min}}) \times R_0^{comp} \quad (9.1)$$

where d_{max} and d_{min} denote the maximum and minimum distance between sensor nodes and the base station respectively, $d(s_i, BS)$ is the distance between the sensor node s_i and the base station, and c is a constant coefficient between 0 and 1. It can

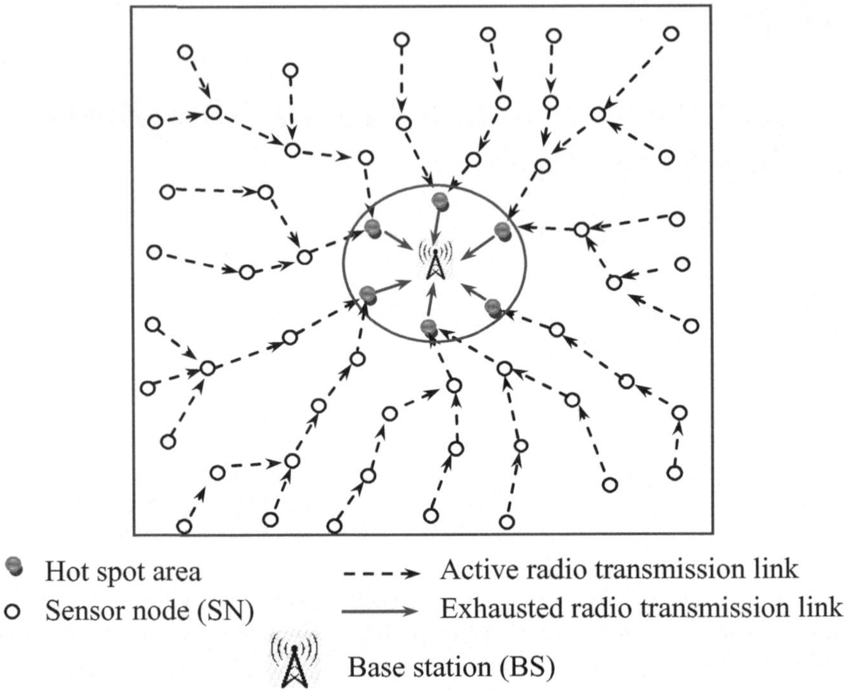

FIGURE 9.1: Network with static base station placed at the geometric center of the field

be observed from the above equation (9.1) that the competition radius varies from $(1-c)R_0^{comp}$ to R_0^{comp}.

The CHs closer to the base station should have smaller cluster size because of the higher energy consumption during the inter-cluster multi-hop routing. Therefore, more clusters should be produced nearer to the base station. This is the rational behind the unequal competition radius of the CHs. The competition radius decreases as the CH's distance to the base station decreases.

All the tentative CHs compete to become the final CH with other CHs within their competition radius. Once a tentative CH finds that its residual energy is more than all the tentative CHs in its competition range, it will win the competition and broadcast a $FINAL\_HEAD\_MSG$ to inform its adjacent tentative CHs which will give up the competition immediately after receiving the message. After selection of CHs, each CH broadcasts a CH_ADV_MSG within its communication rage. Each ordinary sensor node joins its closest CH with the largest received signal strength. The sensor nodes inform the joining message to the CH by sending a $JOIN\_CLUSTER\_MSG$. The organization of intra-cluster data transmission is identical with LEACH. In multi-hop data routing phase, a threshold TD_{MAX} is introduced. If the distance from a CH to the base station is smaller than TD_{MAX}, then the CH directly transmits its aggregated data to the base station. Otherwise the CH should find a relay node to forward its data to the base station.

9.2.2 Multi-hop routing protocol with unequal clustering (MRPUC)

The MRPUC [165] is a multi-hop routing and unequal clustering algorithm. It sets the cluster radius of a sensor node s_i as follows:

$$R_{comp}(s_i) = \frac{d(s_i, BS) \times (R_{max} - R_{min})}{d_{max}} + R_{min} \tag{9.2}$$

where d_{max} is the maximum distance between sensor nodes and BS, and R_{max} and R_{min} are the predefined maximum and minimum cluster radius.

In the cluster setup phase, each node broadcasts a $Hello(ID, E)$ message within the range R_{max} to collect the information of the neighbors. The message consists of node id, ID, and residual energy, E. The sensor node (say s_i) having the largest residual energy among all neighbor nodes is selected as cluster head and broadcasts a $HEAD\_MSG(ID, R_{comp}(s_i), E)$ message. Depending on the received signal strength, each neighbor sensor nodes (say s_j) compute the distance $d(s_i, s_j)$. The normal sensor node (say s_j) joins a cluster head s_k which has the minimum cost which is computed by the following equation:

$$Cost(s_k) = w \times \frac{d(s_j, s_k)}{R_{comp}(s_k)} + (1 - w) \times \frac{(E_{max} - E_k)}{E_{max}} \tag{9.3}$$

where E_{max} is the maximum residual energy of the CHs within the range of s_j, w is the weighted factor for the trade-off between the distance to the CHs and the residual energy of the CHs.

In the data-routing phase, an inter-cluster tree is created where the root of the tree is the base station. A CH s_i chooses its parent node P_i from its set of neighbor CHs, N_i, by the following rule.

$$P_i = \begin{cases} BS & \text{If } d(s_i, BS) < T_{max} \\ s_j & otherwise \end{cases} \tag{9.4}$$

where T_{max} is the threshold transmission range and

$$s_j = Min\{CH\_Cost(s_k), |d(s_k, BS) \leq d(s_i, BS), \forall s_k \in N_i\} \tag{9.5}$$

and

$$CH\_Cost(s_k) = w \times \frac{d^2(s_i, s_k) - d^2(s_k, BS)}{Max\{d^2(s_i, s_k) - d^2(s_k, BS), \forall s_k \in N_i\}} + (1 - w) \times \frac{(E_{max} - E_k)}{E_{max}} \tag{9.6}$$

The above rule says that if the BS is within the communication range T_max, then

BS is the next hop, otherwise the parent node is the minimum-cost node among the neighbor cluster heads that are nearer to the BS than the sender CH. The cost is based on two factors, i.e., energy consumption in data transmission and the residual energy of neighbor cluster heads.

9.2.3 Grid based fault tolerant clustering and routing algorithms

Grid-based fault tolerant clustering and routing algorithms (GFTCRA) [174], divides the whole target area into several grids of equal size. The grids are constructed based on the transmission range of the sensor nodes and each grid forms a cluster in which a normal sensor node is selected as a CH of that cluster. In order to save the energy, all the CHs communicate with the sink via other CHs that are closer to the sink. The algorithm works in round that is divided into following three phases, i.e., clustering phase, routing phase and the fault detection and management phase.

CHs are selected from the normal sensor nodes based on the minimum average distance and energy level which should be above some threshold. In the data routing phase, a CH C_i selects another CH C_j as a relay node which has the minimum distance from the CH C_i. This path is called as primary path of CH C_i. Then the CH C_j follows the same procedure to select the primary path and this process continues until the path reaches the sink. Then, the load of the CHs are balanced based on the residual energy of neighbor CHs.

9.2.4 Unequal multi-hop balanced immune clustering protocol

Unequal multi-hop balanced immune clustering protocol (UMBIC)[175] solves the hot spot problem and balances the energy consumption among the sensor nodes by (1) utilizing the multi-objective immune algorithm (MOIA) to select an optimum number of CHs with more residual energy. Then a routing tree is constructed among the CHs; (2) rotating the role of CH periodically; (3) using the Unequal Clustering Mechanism (UCM) to partition the network into clusters of unequal size; and (4) selecting the best next hop for each CH.

In the cluster building phase, the BS builds optimum clusters and constructs a routing tree among the CHs of these clusters. In the beginning of this phase, the BS calculates the R_C of sensor node s_i based on its distance to the BS and its residual energy as follows:

$$R_C(i) = R_{max} \times \Big(1 - \alpha \times \Big(\frac{d_{maxBS} - d(i, BS)}{d_{maxBS} - d_{minBS}}\Big) - \beta \times \Big(\frac{E_{max} - E_i}{E_{max}}\Big)\Big) \tag{9.7}$$

where d_{maxBS} and d_{minBS} are the longest and shortest distances between the sensor nodes and the BS respectively, $d(i, BS)$ is the distance between sensor node s_i and BS, E_i the residual energy of s_i, E_{max} is the maximum residual energy of sensor nodes, and α and β are constants in the range [0 1] with $\alpha + \beta \leq 1$.

UMBIC protocol utilizes the MOIA to find optimum CHs. Once the CHs are selected, the UMBIC constructs a routing tree among the selected CHs to reduce the transmission energy and to avoid the hot spot problem. The protocol selects the best next hop for each CH based on its distance from the BS, residual energy and the number of hops it takes to reach BS.

9.3 Mobile sink routing (MSR) schemes

Multi-hop communication is an efficient way to save energy of the sensor nodes. However, multi-hop routing using static sink˙ creates the energy hole or hot spot problem[2] (please refer Figure 9.1). It has been observed that the SNs near the sink consumes 98% energy while the SNs farther from the sink consume only 10% of their energy reserves [176]. One technique to stay away from such situations is to utilize the concept of mobility. The SNs can take shifts to be adjoin to the sink because of its mobility, so the routing load is dispersed uniformly among the SNs, and the lifespan of the network can be extended [177]. The other way is to make the sink to roam around the target area and collect the date from the sensor nodes [178]. In this arrangement, the sink is mounted on a vehicle and collect data from the sensor nodes [179]. A MS traverses through the WSN and gathers sensory intelligence per-

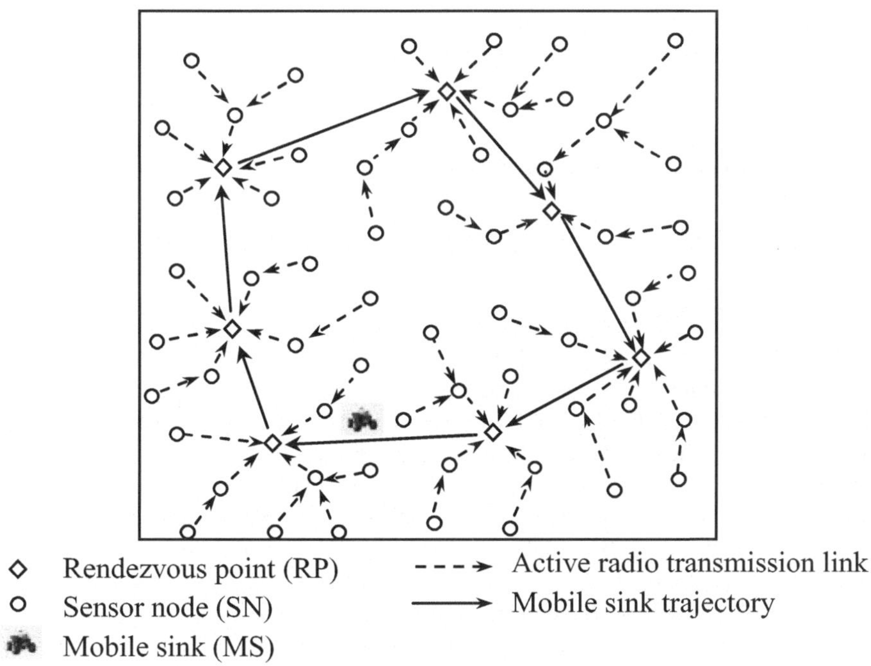

FIGURE 9.2: Mobile data collection in a WSN with mobile sink

sonally (or by implication inside a scant hops) from every SN and afterward takes the gathered information back to its home. As the mobility based technique abolishes the multi-hop transmission, it is anticipated to altogether enhance the system lifetime and the intelligence accumulation rate.

The movement of the sink may be arbitrary [180] or constrained [181]. In the arbitrary movement, the MS going to each SN can have large tour and in the meantime, a few SNs may encounter the problem of memory overflow which outcomes in undesirable information lose. Because of the extended tour stretch, there is a notable latency in information transportation. In the controlled or constrained movement, scientists have suggested rendezvous points (RPs) [182] based MS trajectory. The MS is to halt at some selected sojourn points, i.e., RPs and collect the data from the SNs (refer Figure 9.2) . Therefore, it is a crucial task to outline an effective mobile sink routing (MSR) protocol along with a specific mobile sink tour guaranteeing the scope and the network integration in a WSN. This issue is an NP-hard [119] problem to pick the least aggregate number of RPs with least aggregate physical gap between them.

Utilization of the MS in WSNs is manifested to be immensely worthwhile [119], a portion of the benefits are as per the following. 1) Signal intervention caused by the environmental noise in long range transmission among the SNs is reduced which obliges to the short transmission range betwixt the SNs and sink. 2) The low transmission range outcomes in a remarkable power stockpiling in SNs. 3)the information conveyance over a concise range is thought to be very safe. 4) Besides, utilization of MS gives an improved energy proficiency that results into an extended lifespan of the underlying WSN.

9.3.1 Clustering and routing algorithms

We now present the detailed description of the two methods MobiCluster [127] and cluster-based routing [122] with their illustration in the subsequent sections.

9.3.1.1 MobiCluster

The MobiCluster [127] is an RP based mechanism. Here, the sole belief is that the SNs are installed in civic communities near to the civic conveyance (i.e., bus, train and so on) road. In aforesaid instance, the SNs (which are situated near to the circumference of the underlying area) are recruited as RPs. The key goals of the suggested mechanism are alleviating the total system cost, power consumption related with the long range communication (i.e., multi-hop transmission) and confirming extended network lifetime by utilizing the steady power expenditure amid SNs. These objectives are accomplished by dividing the SNs in clusters or groups and selecting a member SN from every cluster as the cluster head. The member SNs forward there sensory intelligence to specific CHs. At CHs, as the buffered data may have high redundancy, the collected intelligence must be refined. These refined statistics are conveyed to the selected RPs, those are positioned in vicinity of MST and ultimately transmitted to the MS (please refer Figure 9.3). The working of MobiCluster can be

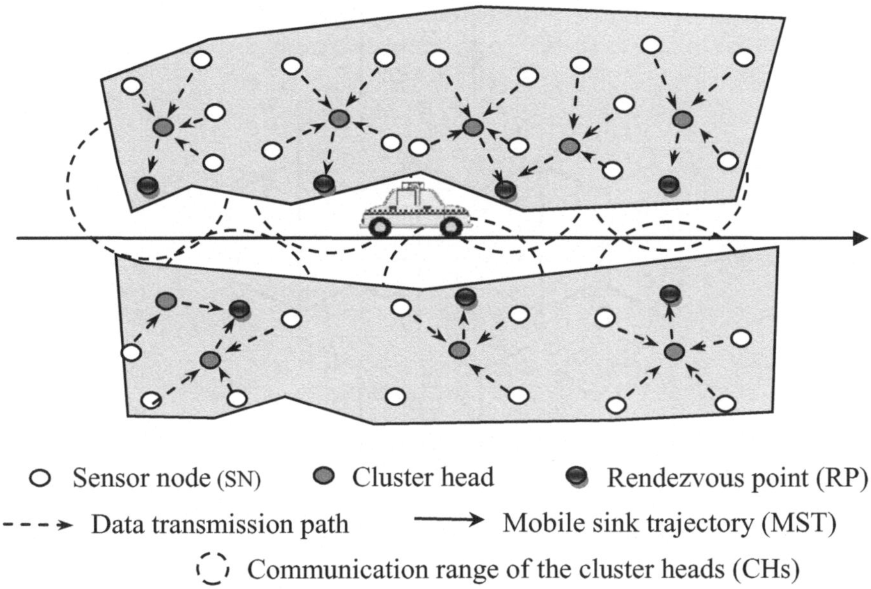

FIGURE 9.3: Physical layout of MobiCluster

divided into five phases.

Clustering phase: The employed clustering protocol utilizes the abstraction of the existing protocol [166] to form the unequal sized clusters. The protocol described in [166] forms multiple clusters of diverse size. The volume of every cluster drops with the increase in distance of its CH from the BS. MobiCluster revises the strategy of this algorithm to form clusters of two diverse volumes subject to physical stretch between the CHs and the MST (mobile sink trajectory). In particular, the nodes situated close to the MST are included in less voluminous clusters and the remaining nodes are included in high voluminous clusters. The MS (e.g., Bus) periodically roves across the sensor field and transmits a BEACON message to the whole SNs. The SNs situated within the MS's transmission range receive and hence they realize that the cluster in their locale will be less voluminous and forwards this intelligence to the other SNs. (please refer Figure 9.4).

Rendezvous points (RPs) selection: When the clustering stage completes, a member SN from each cluster is elected as an RP by the respective CH. This RP must be situated in the transmission range of the MST. The SNs that are situated relatively closer to the MST, accrue comparatively immense amount of BEACON signals and they also have ample amount of power reserves are reasonable SNs to be recruited as RPs. The Authors [25] employ a basic election method by which members of the cluster (nearby the MST) volunteer as RPs. The SNs (serving as RPs) may change

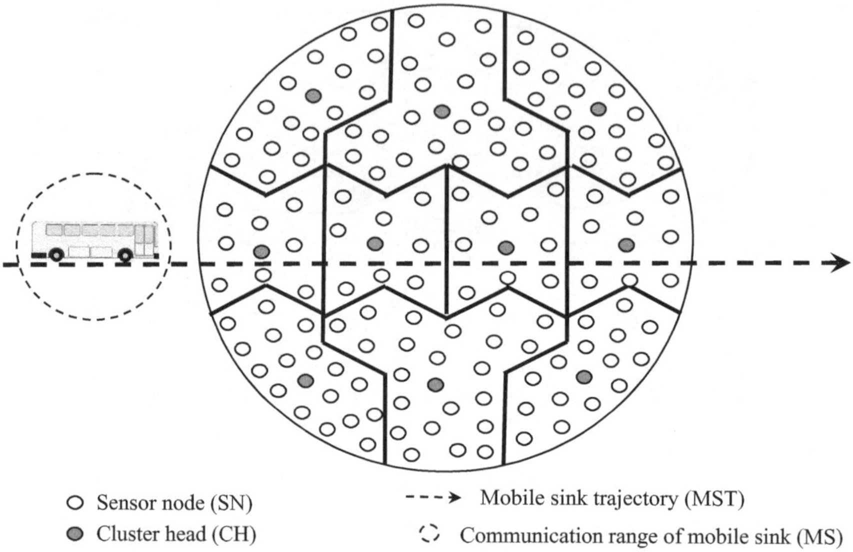

FIGURE 9.4: Clustering phase in MobiCluster

their role with other members form the same cluster, if the energy reserves of a SN falls beneath a certain value. Every SN say (s) manages a counter (cs) to record the number of BEACON signal it has received, whenever a SN listens i^{th} BEACON signal, it restores a "Connection Dropped Timer" (t_{beacon}) fix equivalent to 3 than the counter is increased by one and the time of reception is also registered as (t_i) along with the received BEACON vigor (b_i). It also stores the reception time of the first and utmost received BEACON signals as ($s.t_f$) and ($s.t_l$) respectively.

The *MobiCluster* concludes that the $n - 1$ BEACON signal are dropped or gone payable to the collision at medium access control (MAC) layer and rises the counter via n, only when it has acquired $t_{i+1} \approx t_i + n.t_{beacon}$ where n is greater than one. At the point when the t_{beacon} expires SNs understands that the MS is not in its vicinity anymore and the counter worth is sealed. The moment when the sink has traversed a SN (s), it does not re-visits the same SN until the next tour. Once the counter value is sealed, the proficiency worth (p_w) of a SN is computed depending upon its remaining energy, counter value c_s and the mean of recorded BEACON signals. Afterwards, the SN transmits an RN_Cand_Msg to the respective CH comprising ID of the node, proficiency worth ($s.Prof_{val}$), t_f and t_l to broadcast its nomination to be elected as RP. The SNs having zenith proficiency values are liable to be chosen as RPs. The pseudo code of the algorithm (i.e., performed by every node s after receiving a new BEACON message) is given in Algorithm 13. Where $x_i; i = 1, 2, 3$ are performance factors of remaining energy, value of the counter c_s and the mean signal durability of the acquired BEACON signals respectively. It is worthwhile noting that the line number 1 through 16 of the suggested algorithm are performed only for introductory

Algorithm 13 RN_Candidancy

1: $c_s = 0, s.t_f = 0, s.t_l = 0$
2: Delay for a BEACON signal
3: Store reception time of the BEACON t_1 and its vigor b_1
4: $s.t_f = t_1, s.t_l = t_1, c_s = 1, c_s, r = 1$
5: Begin 'Connection Dropped Timer' t_{beacon}
6: **while** (t_{beacon} is not lapsed) **do**
7:     Delay till t_{beacon} is lapsed or a new BEACON message is received
8:     **if** (a BEACON i is heard) **then**
9:         Store t_i and b_i
10:         $c_s = c_s +$ absolute value of $(t_i - t_i - 1/t_{beacon})$
11:         $c_{s,r} = c_{s,r} + 1$
12:         $s.t_l = t_i$
13:         Restart t_{beacon}
14:     **end if**
15: **end while**
16: $s.pw = x_1.(E_{remaining}/E_{max}) + x_2.c_s + x_3.(\sum^{i=1 \text{ to } c_s} \frac{b_i}{c_s})$
17: Transmit RN_Cand_Msg
18: Stop.

MS tour. Whereas the final (i.e., line number 17) line is performed just after the clustering process.

The competitor SNs from every cluster transmits an RP_Cand_Msg message to its cluster head. In light of the data contained in RN_Cand_Msgs, every CH c selects all the SNs with proficiency worth greater than or equal to a given limit Land stores in set candidate rendezvous point (CRP).For every applicant SN $s_i \in CRP$, interval I_{si} is $[s.t_f, s.t_l]$. At that point the intervals $[s_1.t_f, s_1.t_l]$, $[s2.tf, s_2.t_l]$, p, $[s_k.t_f, s_k.t_l]$ (these intervals represent every SN in CRP) are thought of the vertex of interval graph IG where two vertices v_1 and v_2 are adjacent only if their intervals do not overlap. Likewise, every member I_{si} of the graph IG is provided with a constructive weight i.e., weight $(I_{si}) = s_i.Prof_{pval}$, $1 < i < k$. Afterwards maximum weight subset (MWS)is obtained by employing maximum weight independent set algorithm (MWISA) so that none of two intersects one another. Every member SN of a cluster resembles time interval and the interval which belong to the set MWS are ultimately recruited as RP for respective cluster.

The procedure ensures that the RPs situated inside identical cluster do not participate in data transmission process with one another and will begin conveying their information after the prior finishes. Thus, the radio medium is further utilized, the rate of the collision is reduced and the throughput of the system is escalated. Moreover, *MobiCluster* lowers the interest for intermediate intelligence buffering memory and the data forwarding load is reasonably dispersed throughout the network.

Association of Cluster heads (CHs) to rendezvous points (RPs): It is noteworthy that each selected CH (c) does not have a CRP set. i.e., there is no RP available in the

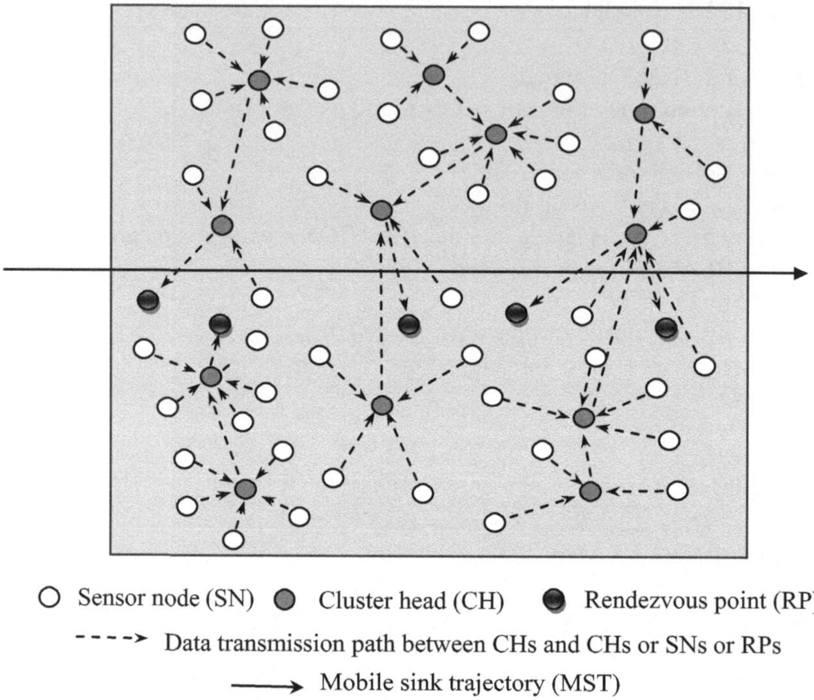

FIGURE 9.5: Derived topology of MobiCluster

communication range of a few CHs positioned far away from the MST. So these CHs are associated with other CHs residing within their communication range. The complete procedure of inter cluster communication can be seen in [127], accessible in the online supporting documentation. *MobiCluster* methodology commonly requires a solitary MS tour to gather (through the reception of BEACON signals) the data required to perform these operations. The Clustering phase begins upon the end of the introductory MS tour. The rendezvous points selection procedure initiates instantly after the Clustering phase. The cluster heads association with rendezvous points trails next. When all three procedures are concluded, every member SN forwards its sensory intelligence to the respective CH by following obeying the constructed inter cluster overlay (please refer Figure 9.5).

Information collection and relaying to the RPs: The MobiCluster convention begins with the intermittent listing of natural information from SNs with a t_r period. The intelligence collected at a SN is relayed to neighboring CH (intra cluster transmission) with a t_c period. Every CH filters the received data as it may have redundancy and transmits the filtered intelligence to other CH or RP. Once the transmitted intelligence reaches at end CH c, it is forwarded to the c's neighboring in a

Algorithm 14 Intelligence_distribution

```
1: d : quantity of buffered intelligence at CH c to be conveyed to RPs.
2: Sort the MWS in decreasing order of proficiency value of RPs.
3: i = 0
4: while (d > 0&&|MWS| > i + 1) do
5:     l_i = s_i.t_f − s_i.t_l and d_i = rate_i.l_i
6:     d_0 = min(d_i, d)
7:     relay(d_0, s_i) /* c relays d_0 intelligence to RP s_i*/
8:     d = d − d_0
9:     i = i+1
10: end while
11: Stop.
```

pipelined manner. For the situation that several RPs are available, intelligence are not evenly dispersed amid them. Rather, the CH encourages the information conveyance by the most appropriate RPs (RPs with the maximum proficiency worth). Intelligence dispersion between RNs should confirm that every RP will have the capacity to shelter its allocated intelligence i.e., all intermediary stored intelligence is delivered and there is no buffer overflow. Thus every CH sorts all the available RP in decreasing order of proficiency value and transmits to every RPsi belong to MWSthe utmost quantity of intelligence (d_i) it can buffer, after removing 'outage prevention allowance'(O). The value of d_i is computed by considering the RP's data rate ($rate_i$)and the duration of the time interval $[s_i.t_f, s_i.t_l]$ which s_i remains in MST's transmission range. The procedure is rehashed for every member RP $s_i \in MWS$ till all available sensory intelligence at c are dispersed amid the RPs. The protocol is performed by every CH c for dispersing the intelligence to the associated RPs and the pseudo code for the protocol is given in Algorithm 14.

Delivery of sensory intelligence from RPs to MS: The final stage of MobiCluster convention includes the conveyance of stored intelligence at RPs to MS. Information conveyance happens along an irregularly accessible connection; consequently, a key necessity is to decide when the passage from a RP to the MS is accessible. Correspondence ought to begin when the passage is accessible and end when the association no more available, i.e., MS is not accepting the data. To mitigate this problem, they utilize an affirmation (acknowledgment) based convention amongst RNs and MS. The MS regularly publicizes a POLL message along its trajectory, reporting its nearness and requesting intelligence as it continues along the way.It is publicized at exact intervals t_{poll} (usually t_{poll} is equivalent to t_{beacon}). RPs gets to know whether the MS is within their communication range or not by utilizing this POLL message. Once a POLL message is received by any CH c, it begins the transmission of buffered towards the MS. The MS affirms every accepted packet of intelligence to the respective RP, hence the RP understands that the association is operative and the intelligence were genuinely conveyed. Then the RP removes the affirmed data from its buffer. It ought to be underscored the enrolment of particular SNs as RPs is liable to altered.

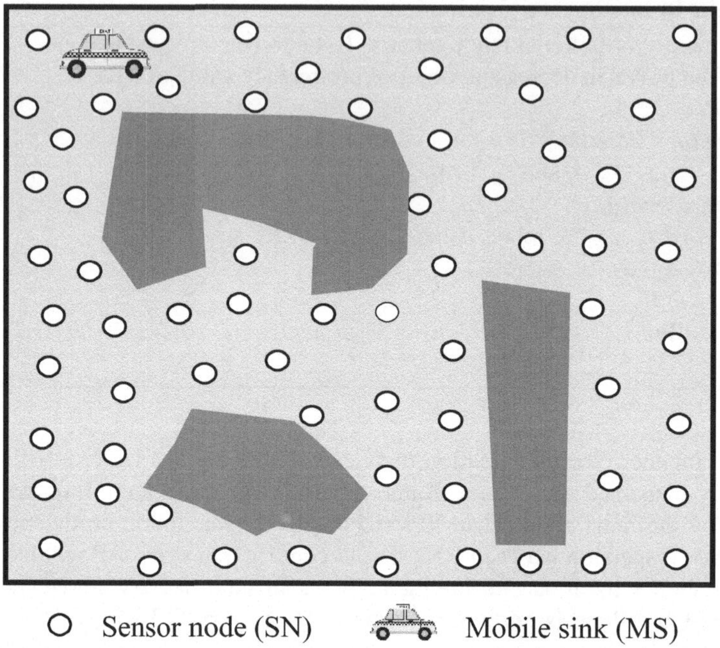

FIGURE 9.6: An illustration of WSN for mobile sink with obstacles

Hence, if the power supply of an RP drops beneath a threshold, it might ask its CH to employ alternative SN as RP in order to boost the network lifetime.To allow RPs replacement, the CH surveys the possible RPs (eliminating the resigning RP) to be educated about their present leftover power reserve and afterward chooses the fresh RP table taking after the strategy of Rendezvous points (RPs) selection.

It is worthwhile to note that in spite of the fact that the clustering procedure acquires thoughts from the protocol of Chen et al. [166] to fabricate a cluster formation (Clustering phase), a completely diverse methodology than that of [166] is taken after for inter cluster routing and intelligence conveyance to the MS. Specifically, new techniques are suggested for selecting proper SNs as RPs, constructing adjustable inter cluster overlay formation, decently dispersing intelligence between RPs and conveying these information to the MS in non-overlapping time frames (Clustering phase via delivery of sensory intelligence from RPs to MS).

9.3.1.2 Cluster based routing

In WSNs, the target area may have a wide range of obstacles. Hence, an algorithm to effectively dispatch the MS with an obstacle-avoiding mobile sink trajectory (OA-MST) is required. This section presents an energy efficient routing convention by utilizing a cluster-based scheme for MS in WSNs [122]. As per the Cluster-based scheme, the SNs are divided into two groups to form clusters and elect CHs, the first

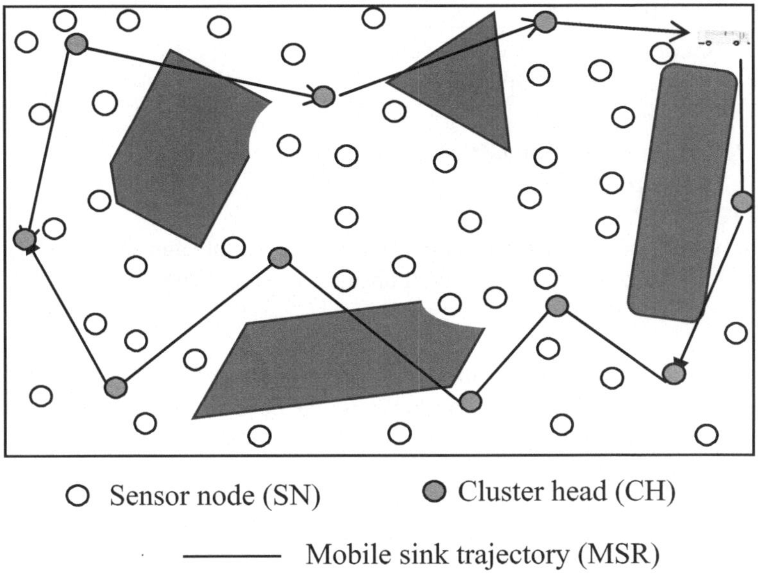

FIGURE 9.7: Partitioning of clusters

group includes the member sensor nodes(MSNs) (nodes which convey the sensory intelligence to CHs) and second group is made of the CHs (nodes to receive and buffer the sensory intelligence originated from MSNs) and the role of CHs are to relay the buffered data to the MS.The MS periodically moves on its trajectory for sensory intelligence collection from the elected CHs in a one hop communication manner. The mobile data collection (MDC) with obstacles in WSNs is a very complex task to determine. To cure this problem, the authors in [122] suggest an effective routing convention using the spanning graphs (SGs). A heuristic tour-planning algorithm depending upon the SG is presented in subsequent parts.

The presented convention gives a grid-based solution towards aforementioned problem where the area under the surveillance is partitioned into the equal size grids. The size of a grid cell is approximately equivalent to the transmission range of the deployed SNs. Since the target area is partitioned into grids, some grid cells will contain the obstacles. When one grid cell possess a portion of obstacle, the convention views the grid cell as an obstacle. Thus, it acquires the appearance of the obstacles so that the discussed problem turns informal.The spanning graph can be built in terms of the appearance of the obstacles. We display some preliminaries required to provide a solution towards the problem stated earlier. A runtime scenario is dissipated in Figure 9.6 where n SNs are randomly installed in a target area with m obstacles.

Cluster partition: To comprehend the MST, the LEACH is utilized (low-energy adaptive clustering hierarchy) convention suggested by Heinzelman et al. [19]. When

Algorithm 15 SpanningGraph

Input: P : set of nodes of pin vertices and obstacle corners.
Output: G : the spanning graph for P.

Sort the set P in non-decreasing order of $x-$ coordinate (P_x)
if ($P_i \subset P_x$ in zone Z2 and Z6 of all obstacle corners) **then**
    EdgeGraph(P_i)
end if
Sort the set P in non-decreasing order of $y-$ coordinate and store it in P_y
if ($P_i \subset P_y$ in zone Z4 and Z8 of all obstacle corners) **then**
    EdgeGraph(P_i)
end if
Sort the set P in non-decreasing order of $x + y-$ coordinate and store it in P_{x+y}
if ($P_i \subset P_{x+y}$ in zone Z1 and Z5 of all obstacle corners and in zone Z1 and Z3 of all pin vertices) **then**
    EdgeGraph(P_i)
end if
Sort the set P in non-decreasing order of $y - x-$coordinate and and store it in P_{y-x}
if ($P_i \subset P_{y-x}$ in zone Z3 and Z7 of all obstacle corners and in zone Z2 and Z4 of all pin vertices) **then**
    EdgeGraph(P_i)
end if
Stop.

CHs are dictated by the LEACH, the MS can move to the nearest CHs to collect intelligence. Henceforth, the most basic energy utilization of SNs is diminish by collecting data with less transmission energy. In Figure 9.7, we dissipate that there are exist ten CHs in single round, which are shown by dark circles. In the event that no obstruction exists, the MS indicated by the bus can move along the line and visit all CHs to gather the sensory intelligence.

Henceforth, the system lifetime can be enhanced outstandingly. Keeping in mind the end goal to balance energy expenditure of SNs, Clusters are re-divided in every data collection round as far as LEACH. See that SNs inside Clusters have the similar opening energy. CHs are chosen depending upon the threshold given by following formula.

$$t(m) = \frac{x}{1 - x \times [k \bmod(\frac{1}{x})]}, \forall m \in S \tag{9.8}$$

where, x is the percentage of CHs, k is the present data collection round number and S is the set of SNs which have not been a CH in the previous $\frac{1}{p}$ data collection rounds. Every SN $m \in S$ picks an arbitrary number q somewhere around 0 and 1 toward the starting of each round. The SN will be chosen as a CH in the present data collection round k if the random number q is less than the threshold.

Algorithm 16 EdgeGraph

Input: The sorted list P with non-decreasing coordinates.
Output: G_s : the edge connection graph for four zone partition.

$U = \Phi$
for ($\forall p_i \in P_s$) **do**
    if (p_i is not corresponding location of blockage b_j) **then**
        add p_i to U
    else
        if (U is not empty) **then**
            $U = \Phi$
            if (b_j is not vertical blockage) **then**
                remove points in four zone of p_i which are from U
            end if
            add these points located in four zone partition of s_i to U_k
            if (U_k not empty) **then**
                select the points m and n from U_k closest to p_i sand p_{i+1}, respectively add edge (s_i, m) and (s_{i+1}, n) to the edge connection graph G_s
            end if
        end if
    end if
end for
Stop.

Heuristic obstacle-avoiding algorithm: In this segment, we introduce a heuristic convention to find a solution to OA-MST. All together to better take care of dispatch problem of MS, the spanning graph algorithm is utilized to construct all the obstacle-avoiding paths for MS in a WSN with obstacles. Besides, one of the obstacle-avoiding paths is used as OA-MST. The spanning graph algorithm is given as follows.

Spanning graph algorithm: Basically, the obstacle-avoiding mobile sink trajectory (OA-MST) problem is related to the Traveling Salesman Problem (TSP) which is an established issue now. To solve the TSP problem the minimum spanning tree algorithm can be utilized. Hence, we use the minimum spanning tree algorithm to determine OA-MST in a WSN. Here, a SG is an undirected graph that comprises every minimum spanning tree and we will talk about how to develop the SG. A few literatures have suggested the spanning graph construction (SPC) [183]. We use the sweep line algorithm to develop the SG [183]. The obstacle - avoiding spanning graph (OA-

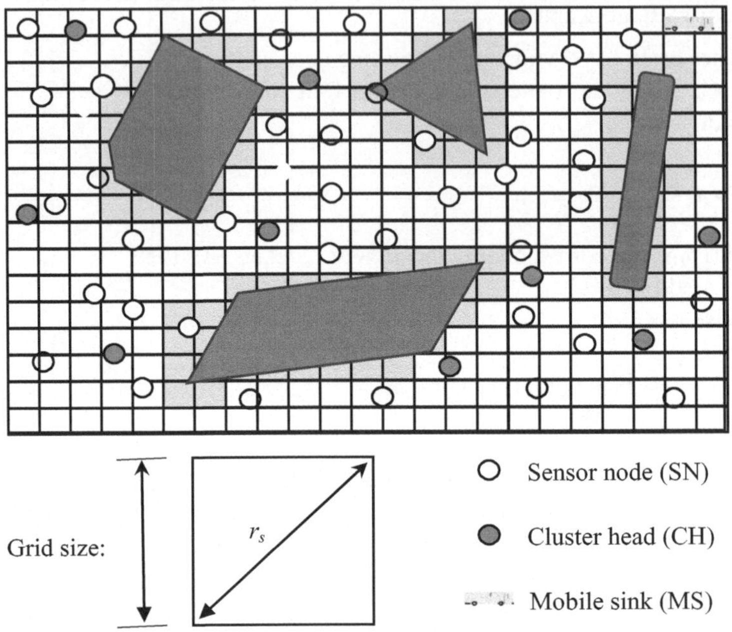

FIGURE 9.8: Grid-graph of target area

SG) is the arrangement of edges that can be framed by making associations amongst terminals and obstacle niches. Once an SG is developed, the limitless conceivable destinations for the MS traversal will be decreased to a restricted set of locales. In this manner, the convention based on the SG forms it more effective to schedule for the MS. The pseudo-code for SpanningGraph and EdgeGraph is given in Algorithm 15 and Algorithm 16 respectively.

The grid division is given in Figure 9.8 in target area. Alluding to Figure 9.8, grey regions of the target area are also assumed as obstacles where the MS cannot travel. The MS scheduling is simplified by considering that there is single location within every grid cell. MS visits these locations to collect sensory intelligence form CH inside the grid cell. The MS location demonstrated by the red circle is normally positioned in geometric center of the grid cell. Hence, we acquire regularization formation of obstacles by utilizing mentioned technique.

Obstacle-avoiding spanning graph construction:

Till now, we have depicted the cluster partition algorithm where SNs are divided into groups to form cluster and select CHs from member SNs. Using sink's locations relative to CHs and locations of obstacle corners, we can build SG for the OA-MST by utilizing above algorithm. Figure 9.9 introduces the OA-MST of Figure 9.8, where MS travels to every CHs to collect sensory intelligence to collect sensory intelligence along its MST. Here, we utilize grid based conventions to OA-MST. The target area is

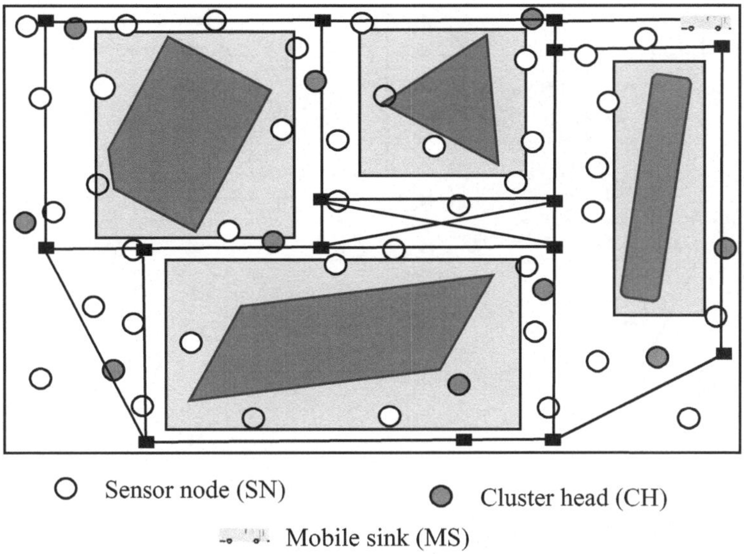

FIGURE 9.9: Spanning graph of target area

partitioned into equal size grid cells by utilizing the grid-based conventions. Clearly, obstacles' edges bisect grid cells and grid cells may contain obstacles. Along these lines, we get regularization state of obstacles which avails an easier OA-MST construction.

Let the transmission range and sensing range of every SN be r_c and r_s correspondingly. It is shown [15] that if there exist no hole in the target area, the grid size r will be $\min(r_s/2, r_c/2)$. Here, we suppose that the transmission range rc of every SN lengthy adequate. Thus, the gird size r is picked to be $r_s/sqrt(2)$ so that no hole available in the target area. As per the SG of the target area in Figure 9.9, we can eventually obtain OA-MST. Fundamentally, the OA-MST is same to the TSP that is a traditional issue.

It has been demonstrated that the TSP is a classical NP-Complete problem. Because of the unpredictability of the TSP, we embrace an approximate convention to take care of the scheduling problem of the MS. The primary presumptions with respect to the TSP is that the input is a complete graph (CG). In any case, the SG that we have built is not a CG. Here, we can without much of a stretch use the Warshall-Floyd procedure to develop the CG as far by means of the SG. In light of the CG, we can undoubtedly obtain the OA-MST from by the concept of two parts. Figure 9.10 demonstrates the obstacle-avoiding mobile sink trajectory (OA-MST) obtained by cluster-based routing scheme.

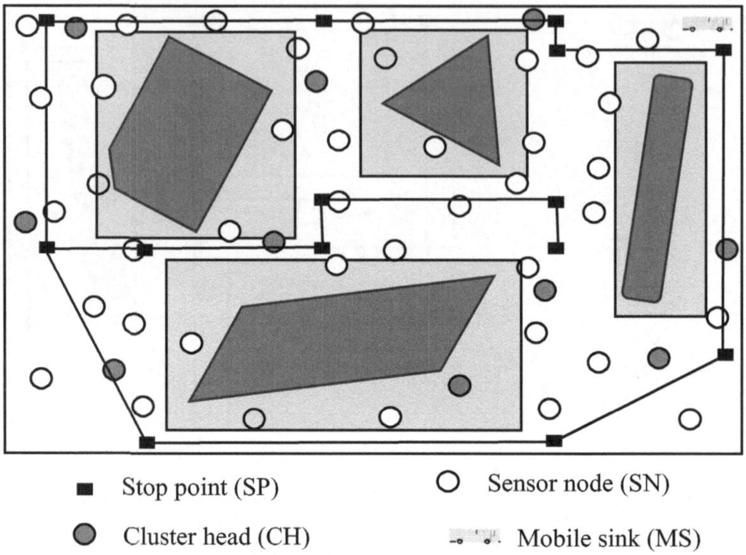

FIGURE 9.10: Obtained OA-MST

9.4 Chapter summary

This chapter mainly focuses on clustering and routing algorithms based on hot-spot problem. Unequal clustering and mobile sink based routing algorithms are discussed. Some well-known unequal clustering algorithms like EEUC, MRPUC, GFTCRA and UMBIC are discussed. In mobile sink based routing, the detailed description of the two methods MobiCluster and Cluster-based routing with their illustration are presented.

Bibliography

[1] Ian F. Akyildiz and Mehmet Can Vuran. *Wireless sensor networks*, volume 4. John Wiley & Sons, 2010.

[2] Ian F. Akyildiz, Weilian Su, Yogesh Sankarasubramaniam, and Erdal Cayirci. Wireless sensor networks: A survey. *Computer networks*, 38(4):393–422, 2002.

[3] Jennifer Yick, Biswanath Mukherjee, and Dipak Ghosal. Wireless sensor network survey. *Computer networks*, 52(12):2292–2330, 2008.

[4] Giuseppe Anastasi, Marco Conti, Mario Di Francesco, and Andrea Passarella. Energy conservation in wireless sensor networks: A survey. *Ad Hoc Networks*, 7(3):537–568, 2009.

[5] Emanuele Lattanzi, Edoardo Regini, Andrea Acquaviva, and Alessandro Bogliolo. Energetic sustainability of routing algorithms for energy-harvesting wireless sensor networks. *Computer Communications*, 30(14):2976–2986, 2007.

[6] Benton H. Calhoun, Denis C. Daly, Naveen Verma, Daniel F. Finchelstein, David D. Wentzloff, Alice Wang, S. Cho, and Anantha P. Chandrakasan. Design considerations for ultra-low energy wireless microsensor nodes. *Computers, IEEE Transactions on*, 54(6):727–740, 2005.

[7] Yong-hong Zeng, Xue-cheng Zou, Zheng-lin Liu, and Jian-ming Lei. Low-power clock-less hardware implementation of the rijndael s-box for wireless sensor networks. *The Journal of China Universities of Posts and Telecommunications*, 14(4):104–109, 2007.

[8] Ahmad Abed Alhameed Alkhatib and Gurvinder Singh Baicher. MAC layer overview for wireless sensor networks. In *Int. Conf. on Computer Networks and Communication Systems*, volume 35, pages 16–19, 2012.

[9] M. Aykut Yigitel, Ozlem Durmaz Incel, and Cem Ersoy. QoS-aware MAC protocols for wireless sensor networks: A survey. *Computer Networks*, 55(8):1982–2004, 2011.

[10] Ilker Demirkol, Cem Ersoy, and Fatih Alagoz. MAC protocols for wireless sensor networks: A survey. *Communications Magazine, IEEE*, 44(4):115–121, 2006.

[11] Rodolfo de Paz Alberola and Dirk Pesch. Duty cycle learning algorithm (DCLA) for IEEE 802.15. 4 beacon-enabled wireless sensor networks. *Ad Hoc Networks*, 10(4):664–679, 2012.

[12] Ricardo C. Carrano, Diego Passos, Luiz C.S. Magalhães, and Célio V.N. Albuquerque. A comprehensive analysis on the use of schedule-based asynchronous duty cycling in wireless sensor networks. *Ad Hoc Networks*, 16:142–164, 2014.

[13] Roja Chandanala, Wei Zhang, Radu Stoleru, and Myounggyu Won. On combining network coding with duty-cycling in flood-based wireless sensor networks. *Ad Hoc Networks*, 11(1):490–507, 2013.

[14] Maher Rebai, Matthieu Le Berre, Hichem Snoussi, Faicel Hnaien, and Lyes Khoukhi. Sensor deployment optimization methods to achieve both coverage and connectivity in wireless sensor networks. *Computers & Operations Research*, 59:11–21, 2015.

[15] Shad Roundy, Paul Kenneth Wright, and Jan M Rabaey. *Energy scavenging for wireless sensor networks*. Springer, 2003.

[16] Gregory J. Pottie and William J. Kaiser. Wireless integrated network sensors. *Communications of the ACM*, 43(5):51–58, 2000.

[17] Crossbow micaz mote specifications. http://www.xbow.com.

[18] Wendi Beth Heinzelman. *Application-specific protocol architectures for wireless networks*. PhD thesis, Massachusetts Institute of Technology, 2000.

[19] Wendi B. Heinzelman, Anantha P. Chandrakasan, and Hari Balakrishnan. An application-specific protocol architecture for wireless microsensor networks. *Wireless Communications, IEEE Transactions on*, 1(4):660–670, 2002.

[20] Isabel Dietrich and Falko Dressler. On the lifetime of wireless sensor networks. *ACM Transactions on Sensor Networks (TOSN)*, 5(1):5, 2009.

[21] Ritesh Madan, Shuguang Cui, Sanjay Lall, and Andrea Goldsmith. Cross-layer design for lifetime maximization in interference-limited wireless sensor networks. *Wireless Communications, IEEE Transactions on*, 5(11):3142–3152, 2006.

[22] Mihaela Cardei and Ding-Zhu Du. Improving wireless sensor network lifetime through power aware organization. *Wireless Networks*, 11(3):333–340, 2005.

[23] Pan J., Hou Y. T., Cai L., Shi Y., and Shen S. X. Topology control for wireless sensor networks. In *Proceedings of the Ninth Annual International Conference on Mobile Computing and Networking*, pages 286–299, 2003.

[24] Lilia Paradis and Qi Han. A survey of fault management in wireless sensor networks. *Journal of Network and Systems Management*, 15(2):171–190, 2007.

[25] Marvin Rausand and Arnljot Høyland. *System reliability theory: models, statistical methods, and applications*, volume 396. John Wiley & Sons, 2004.

[26] Jae-Joon Lee, Bhaskar Krishnamachari, and C.C. Jay Kuo. Aging analysis in large-scale wireless sensor networks. *Ad Hoc Networks*, 6(7):1117–1133, 2008.

[27] Pratyay Kuila and Prasanta K. Jana. Energy efficient clustering and routing algorithms for wireless sensor networks: Particle swarm optimization approach. *Engineering Applications of Artificial Intelligence*, 33:127–140, 2014.

[28] Pratyay Kuila and Prasanta K. Jana. A novel differential evolution based clustering algorithm for wireless sensor networks. *Applied Soft Computing*, 25:414–425, 2014.

[29] Pratyay Kuila, Suneet Kumar Gupta, and Prasanta K. Jana. A novel evolutionary approach for load balanced clustering problem for wireless sensor networks. *Swarm and Evolutionary Computation*, 12:48–56, 2013.

[30] Suneet Kumar Gupta, Pratyay Kuila, and Prasanta K. Jana. GAR: An energy efficient GA-based routing for wireless sensor networks. In *International Conference on Distributed Computing and Internet Technology 2013, LNCS (Springer)*, volume 7753, pages 267–277. Springer, 2013.

[31] Ataul Bari, Shamsul Wazed, Arunita Jaekel, and Subir Bandyopadhyay. A genetic algorithm based approach for energy efficient routing in two-tiered sensor networks. *Ad Hoc Networks*, 7(4):665–676, 2009.

[32] Chor Ping Low, Can Fang, Jim Mee Ng, and Yew Hock Ang. Efficient load-balanced clustering algorithms for wireless sensor networks. *Computer Communications*, 31(4):750–759, 2008.

[33] Ataul Bari, Arunita Jaekel, and Subir Bandyopadhyay. Clustering strategies for improving the lifetime of two-tiered sensor networks. *Computer Communications*, 31(14):3451–3459, 2008.

[34] Xiuzhen Cheng, Ding-Zhu Du, Lusheng Wang, and Baogang Xu. Relay sensor placement in wireless sensor networks. *Wireless Networks*, 14(3):347–355, 2008.

[35] Ataul Bari, Arunita Jaekel, and Subir Bandyopadhyay. Maximizing the lifetime of two-tiered sensor networks. In *Electro/information Technology, 2006 IEEE International Conference on*, pages 222–226. IEEE, 2006.

[36] Jian Tang, Bin Hao, and Arunabha Sen. Relay node placement in large scale wireless sensor networks. *Computer communications*, 29(4):490–501, 2006.

[37] Emil Falck, Patrik Floréen, Petteri Kaski, Jukka Kohonen, and Pekka Orponen. Balanced data gathering in energy-constrained sensor networks. In *Algorithmic Aspects of Wireless Sensor Networks*, pages 59–70. Springer, 2004.

[38] Gaurav Gupta and Mohamed Younis. Load-balanced clustering of wireless sensor networks. In *Communications, 2003. ICC'03. IEEE International Conference on*, volume 3, pages 1848–1852. IEEE, 2003.

[39] Gaurav Gupta and Mohamed Younis. Fault-tolerant clustering of wireless sensor networks. In *Wireless Communications and Networking, 2003. WCNC 2003. 2003 IEEE*, volume 3, pages 1579–1584. IEEE, 2003.

[40] Habib M. Ammari and Sajal K. Das. A trade-off between energy and delay in data dissemination for wireless sensor networks using transmission range slicing. *Computer Communications*, 31(9):1687–1704, 2008.

[41] Y. Thomas Hou, Yi Shi, Jianping Pan, and Scott F. Midkiff. Lifetime-optimal data routing in wireless sensor networks without flow splitting. In *Workshop on Broadband Advanced Sensor Networks, San Jose, CA*, 2004.

[42] WATERSENSE project consortium. available online:http://www.projectwatersense.nl. 2010.

[43] Mohamed Amine Kafi, Yacine Challal, Djamel Djenouri, Messaoud Doudou, Abdelmadjid Bouabdallah, and Nadjib Badache. A study of wireless sensor networks for urban traffic monitoring: applications and architectures. *Procedia computer science*, 19:617–626, 2013.

[44] Milica Pejanović Đurišić, Zhilbert Tafa, Goran Dimić, and Veljko Milutinović. A survey of military applications of wireless sensor networks. In *2012 Mediterranean conference on embedded computing (MECO)*, pages 196–199. IEEE, 2012.

[45] Oladayo Bello, Sherali Zeadally, and Mohammad Badra. Network layer inter-operation of device-to-device communication technologies in internet of things (iot). *Ad Hoc Networks*, 2016.

[46] C.P. Kruger, A.M. Abu-Mahfouz, and G.P. Hancke. Rapid prototyping of a wireless sensor network gateway for the internet of things using off-the-shelf components. In *Industrial Technology (ICIT), 2015 IEEE International Conference on*, pages 1926–1931. IEEE, 2015.

[47] Wendi Rabiner Heinzelman, Amit Sinha, Alice Wang, and Anantha P. Chandrakasan. Energy-scalable algorithms and protocols for wireless microsensor networks. In *Acoustics, Speech, and Signal Processing, 2000. ICASSP'00. Proceedings. 2000 IEEE International Conference on*, volume 6, pages 3722–3725. IEEE, 2000.

[48] Rex Min, Manish Bhardwaj, S Cho, Amit Sinha, Eugene Shih, Alice Wang, and Anantha Chandrakasan. An architecture for a power-aware distributed microsensor node. In *Signal Processing Systems, 2000. SiPS 2000. 2000 IEEE Workshop on*, pages 581–590. IEEE, 2000.

[49] Alec Woo and David E Culler. A transmission control scheme for media access in sensor networks. In *Proceedings of the 7th annual international conference on Mobile computing and networking*, pages 221–235. ACM, 2001.

[50] Wei Ye, John Heidemann, and Deborah Estrin. An energy-efficient MAC protocol for wireless sensor networks. In *INFOCOM 2002. Twenty-First Annual Joint Conference of the IEEE Computer and Communications Societies. Proceedings. IEEE*, volume 3, pages 1567–1576. IEEE, 2002.

[51] Eugene Shih, Seong-Hwan Cho, Nathan Ickes, Rex Min, Amit Sinha, Alice Wang, and Anantha Chandrakasan. Physical layer driven protocol and algorithm design for energy-efficient wireless sensor networks. In *Proceedings of the 7th annual international conference on Mobile computing and networking*, pages 272–287. ACM, 2001.

[52] Kemal Akkaya and Mohamed Younis. A survey on routing protocols for wireless sensor networks. *Ad hoc networks*, 3(3):325–349, 2005.

[53] Olutayo Boyinbode, Hanh Le, and Makoto Takizawa. A survey on clustering algorithms for wireless sensor networks. *International Journal of Space-Based and Situated Computing*, 1(2):130–136, 2011.

[54] Ameer Ahmed Abbasi and Mohamed Younis. A survey on clustering algorithms for wireless sensor networks. *Computer communications*, 30(14):2826–2841, 2007.

[55] Jamal N Al-Karaki and Ahmed E Kamal. Routing techniques in wireless sensor networks: A survey. *Wireless Communications, IEEE*, 11(6):6–28, 2004.

[56] Congfeng Jiang, Daomin Yuan, and Yinghui Zhao. Towards clustering algorithms in wireless sensor networks - a survey. In *Wireless Communications and Networking Conference, 2009. WCNC 2009. IEEE*, pages 1–6. IEEE, 2009.

[57] Wenjing Guo and Wei Zhang. A survey on intelligent routing protocols in wireless sensor networks. *Journal of Network and Computer Applications*, 38:185–201, 2014.

[58] Adamu Murtala Zungeru, Li-Minn Ang, and Kah Phooi Seng. Classical and swarm intelligence based routing protocols for wireless sensor networks: A survey and comparison. *Journal of Network and Computer Applications*, 35(5):1508–1536, 2012.

[59] Priyantha Kumarawadu, Dan J. Dechene, Marco Luccini, and Allan Sauer. Algorithms for node clustering in wireless sensor networks: A survey. In *Information and Automation for Sustainability, 2008. ICIAFS 2008. 4th International Conference on*, pages 295–300. IEEE, 2008.

[60] S.R. Boselin Prabhu and S. Sophia. A survey of adaptive distributed clustering algorithms for wireless sensor networks. *International Journal of Computer Science and Engineering Survey*, 2(4):165–176, 2011.

[61] Gaurav Gupta and Mohamed Younis. Performance evaluation of load-balanced clustering of wireless sensor networks. In *Telecommunications, 2003. ICT 2003. 10th international conference on*, volume 2, pages 1577–1583. IEEE, 2003.

[62] A. Tarachand, Vikas Kumar, Abhishek Raj, Ashish Kumar, and Prasanta K. Jana. An energy efficient load balancing algorithm for cluster-based wireless sensor networks. In *India Conference (INDICON), 2012 Annual IEEE*, pages 1250–1254. IEEE, 2012.

[63] Haifeng Jiang, Jiansheng Qian, and Jie Zhao. Cluster head load balanced clustering routing protocol for wireless sensor networks. In *Mechatronics and Automation, 2009. ICMA 2009. International Conference on*, pages 4002–4006. IEEE, 2009.

[64] Yaping Deng and Yaming Hu. A load balance clustering algorithm for heterogeneous wireless sensor networks. In *E-Product E-Service and E-Entertainment (ICEEE), 2010 International Conference on*, pages 1–4. IEEE, 2010.

[65] Yihui Li, Gaoxi Xiao, Gurpreet Singh, and Rashmi Gupta. Algorithms for finding best locations of cluster heads for minimizing energy consumption in wireless sensor networks. *Wireless networks*, 19(7):1755–1768, 2013.

[66] Pratyay Kuila and Prasanta K. Jana. Evolutionary computing approaches for clustering and routing in wireless sensor networks. In *Handbook of Research on Natural Computing for Optimization Problems*, pages 246–266. IGI Global, 2016.

[67] Sajid Hussain, Abdul Wasey Matin, and Obidul Islam. Genetic algorithm for hierarchical wireless sensor networks. *Journal of Networks*, 2(5), 2007.

[68] Enan A Khalil and BaraA Attea. Energy-aware evolutionary routing protocol for dynamic clustering of wireless sensor networks. *Swarm and Evolutionary Computation*, 1(4):195–203, 2011.

[69] Shiyuan Jin, Ming Zhou, and Annie S Wu. Sensor network optimization using a genetic algorithm. In *Proceedings of the 7th World Multiconference on Systemics, Cybernetics and Informatics*, pages 109–116, 2003.

[70] S. Mudundi and Hesham H. Ali. A new robust genetic algorithm for dynamic cluster formation in wireless sensor networks. *Proceedings of Wireless and Optical Communications, Montreal, Quebec, Canada*, 2007.

[71] Xin-She Yang. Harmony search as a metaheuristic algorithm. In *Music-inspired harmony search algorithm*, pages 1–14. Springer, 2009.

[72] Moslem Afrashteh Mehr. Design and implementation a new energy efficient clustering algorithm using genetic algorithm for wireless sensor networks. *Engineering and Technology*, 53(2):430–433, 2011.

[73] Xiaoyan Cui and Zhao Liu. BCEE: a balanced-clustering, energy-efficient hierarchical routing protocol in wireless sensor networks. In *Network Infrastructure and Digital Content, 2009. IC-NIDC 2009. IEEE International Conference on*, pages 26–30. IEEE, 2009.

[74] Wang Guifeng, Wang Yong, and Tao Xiaoling. An ant colony clustering routing algorithm for wireless sensor networks. In *Genetic and Evolutionary Computing, 2009. WGEC'09. 3rd International Conference on*, pages 670–673. IEEE, 2009.

[75] Uday K. Chakraborty, Sajal K. Das, and Travis E. Abbott. Energy-efficient routing in hierarchical wireless sensor networks using differential-evolution-based memetic algorithm. In *Evolutionary Computation (CEC), 2012 IEEE Congress on*, pages 1–8. IEEE, 2012.

[76] N.M. Abdul Latiff, C.C. Tsimenidis, and B.S. Sharif. Energy-aware clustering for wireless sensor networks using particle swarm optimization. In *Personal, Indoor and Mobile Radio Communications, 2007. PIMRC 2007. IEEE 18th International Symposium on*, pages 1–5. IEEE, 2007.

[77] Buddha Singh and D.K. Lobiyal. Energy-aware cluster head selection using particle swarm optimization and analysis of packet retransmissions in WSN. *Procedia Technology*, 4:171–176, 2012.

[78] Muhammad Saleem, Gianni A. Di Caro, and Muddassar Farooq. Swarm intelligence based routing protocol for wireless sensor networks: Survey and future directions. *Information Sciences*, 181(20):4597–4624, 2011.

[79] Raghavendra V. Kulkarni and Ganesh Kumar Venayagamoorthy. Particle swarm optimization in wireless-sensor networks: A brief survey. *Systems, Man, and Cybernetics, Part C: Applications and Reviews, IEEE Transactions on*, 41(2):262–267, 2011.

[80] Xuxun Liu and Desi He. Ant colony optimization with greedy migration mechanism for node deployment in wireless sensor networks. *Journal of Network and Computer Applications*, 39:310–318, 2014.

[81] Chi Lin, Guowei Wu, Feng Xia, Mingchu Li, Lin Yao, and Zhongyi Pei. Energy efficient ant colony algorithms for data aggregation in wireless sensor networks. *Journal of Computer and System Sciences*, 78(6):1686–1702, 2012.

[82] Wen-Hwa Liao, Yucheng Kao, and Ru-Ting Wu. Ant colony optimization based sensor deployment protocol for wireless sensor networks. *Expert Systems with Applications*, 38(6):6599–6605, 2011.

[83] Nuo Wei, Qiang Guo, Ming-Lei Shu, Jia-liang LÜ, and Ming Yang. Three-dimensional localization algorithm of wireless sensor networks base on particle swarm optimization. *The Journal of China Universities of Posts and Telecommunications*, 19:7–12, 2012.

[84] Adamu Murtala Zungeru, Li-Minn Ang, and Kah Phooi Seng. Termite-hill: Performance optimized swarm intelligence based routing algorithm for wireless sensor networks. *Journal of Network and Computer Applications*, 35(6):1901–1917, 2012.

[85] Orhan Dengiz, Abdullah Konak, and Alice E. Smith. Connectivity management in mobile ad hoc networks using particle swarm optimization. *Ad Hoc Networks*, 9(7):1312–1326, 2011.

[86] Wendi Rabiner Heinzelman, Joanna Kulik, and Hari Balakrishnan. Adaptive protocols for information dissemination in wireless sensor networks. In *Proceedings of the 5th annual ACM/IEEE international conference on Mobile computing and networking*, pages 174–185. ACM, 1999.

[87] S. Hedetniemi and A. Liestman. A survey of gossiping and broadcasting in communication networks. *Networks*, 18(4):319–349, 1988.

[88] Chalermek Intanagonwiwat, Ramesh Govindan, and Deborah Estrin. Directed diffusion: a scalable and robust communication paradigm for sensor networks. In *Proceedings of the 6th annual international conference on Mobile computing and networking*, pages 56–67. ACM, 2000.

[89] Deborah Estrin, Ramesh Govindan, John Heidemann, and Satish Kumar. Next century challenges: Scalable coordination in sensor networks. In *Proceedings of the 5th annual ACM/IEEE international conference on Mobile computing and networking*, pages 263–270. ACM, 1999.

[90] Sudhanshu Tyagi and Neeraj Kumar. A systematic review on clustering and routing techniques based upon LEACH protocol for wireless sensor networks. *Journal of Network and Computer Applications*, 36(2):623–645, 2013.

[91] V. Geetha, P.V. Kallapur, and Sushma Tellajeera. Clustering in wireless sensor networks: Performance comparison of LEACH & LEACH-C protocols using NS2. *Procedia Technology*, 4:163–170, 2012.

[92] Mohammad Masdari, Sadegh Mohammadzadeh Bazarchi, and Moazam Bidaki. Analysis of secure LEACH-based clustering protocols in wireless sensor networks. *Journal of Network and Computer Applications*, 36(4):1243–1260, 2013.

[93] Aimin Wang, Dailiang Yang, and Dayang Sun. A clustering algorithm based on energy information and cluster heads expectation for wireless sensor networks. *Computers & Electrical Engineering*, 38(3):662–671, 2012.

[94] Abbas Nayebi and Hamid Sarbazi-Azad. Performance modeling of the LEACH protocol for mobile wireless sensor networks. *Journal of parallel and distributed computing*, 71(6):812–821, 2011.

[95] Hakan Bagci and Adnan Yazici. An energy aware fuzzy approach to unequal clustering in wireless sensor networks. *Applied Soft Computing*, 13(4):1741–1749, 2013.

[96] Hasan Al-Refai, Ali Al-Awneh, Khaldoun Batiha, Amer Abu Ali, and Yehia M. El Rahman. Efficient routing LEACH (ER-LEACH) enhanced on LEACH protocol in wireless sensor networks. *International Journal of Academic Research*, 3(3), 2011.

[97] Otgonchimeg Buyanjargal and Youngmi Kwon. AEEC-Adaptive and energy efficient clustering algorithm for content based wireless sensor networks. In *Computer Science and its Applications, 2009. CSA'09. 2nd International Conference on*, pages 1–6. IEEE, 2009.

[98] Stephanie Lindsey, Cauligi Raghavendra, and Krishna M. Sivalingam. Data gathering algorithms in sensor networks using energy metrics. *Parallel and Distributed Systems, IEEE Transactions on*, 13(9):924–935, 2002.

[99] Stephanie Lindsey and Cauligi S. Raghavendra. PEGASIS: Power-efficient gathering in sensor information systems. In *Aerospace conference proceedings, 2002. IEEE*, volume 3, pages 3–1125. IEEE, 2002.

[100] Ossama Younis and Sonia Fahmy. HEED: A hybrid, energy-efficient, distributed clustering approach for ad hoc sensor networks. *Mobile Computing, IEEE Transactions on*, 3(4):366–379, 2004.

[101] Arati Manjeshwar and Dharma P. Agrawal. TEEN: a routing protocol for enhanced efficiency in wireless sensor networks. In *Proceedings of the International Parallel & Distributed Processing Symposium*, volume 1, page 189, 2001.

[102] Arati Manjeshwar and Dharma P Agrawal. APTEEN: a hybrid protocol for efficient routing and comprehensive information retrieval in wireless sensor networks. In *Proceedings of the International Parallel & Distributed Processing Symposium*, volume 2, page 48, 2002.

[103] Li Qing, Qingxin Zhu, and Mingwen Wang. Design of a distributed energy-efficient clustering algorithm for heterogeneous wireless sensor networks. *Computer communications*, 29(12):2230–2237, 2006.

[104] Chang-Soo Ok, Seokcheon Lee, Prasenjit Mitra, and Soundar Kumara. Distributed energy balanced routing for wireless sensor networks. *Computers & Industrial Engineering*, 57(1):125–135, 2009.

[105] Anfeng Liu, Ju Ren, Xu Li, Zhigang Chen, and Xuemin Sherman Shen. Design principles and improvement of cost function based energy aware routing algorithms for wireless sensor networks. *Computer Networks*, 56(7):1951–1967, 2012.

[106] Tao Liu, Qingrui Li, and Ping Liang. An energy-balancing clustering approach for gradient-based routing in wireless sensor networks. *Computer Communications*, 35(17):2150–2161, 2012.

[107] Eduardo Cañete, Manuel Díaz, Luis Llopis, and Bartolomé Rubio. HERO: A hierarchical, efficient and reliable routing protocol for wireless sensor and actor networks. *Computer Communications*, 35(11):1392–1409, 2012.

[108] Carlos Abreu, Manuel Ricardo, and P.M. Mendes. Energy-aware routing for biomedical wireless sensor networks. *Journal of Network and Computer Applications*, 2014.

[109] Chenn-Jung Huang, Yu-Wu Wang, Hsiu-Hui Liao, Chin-Fa Lin, Kai-Wen Hu, and Tun-Yu Chang. A power-efficient routing protocol for underwater wireless sensor networks. *Applied Soft Computing*, 11(2):2348–2355, 2011.

[110] Zhi-yuan Li and Ru-chuan Wang. Load balancing-based hierarchical routing algorithm for wireless multimedia sensor networks. *The Journal of China Universities of Posts and Telecommunications*, 17:51–59, 2010.

[111] Zhen Hong, Li Yu, and Gui-Jun Zhang. Efficient and dynamic clustering scheme for heterogeneous multi-level wireless sensor networks. *Acta Automatica Sinica*, 39(4):454–460, 2013.

[112] Suneet Kumar Gupta, Pratyay Kuila, and Prasanta K. Jana. E3BFT: Energy efficient and energy balanced fault tolerance clustering in wireless sensor networks. In *Contemporary Computing and Informatics (IC3I), 2014 International Conference on*, pages 714–719. IEEE, 2014.

[113] Shu-jie Guo, Jie Zheng, Yu-gui Qu, Bao-Hua Zhao, and Quan-Ke Pan. Clustering and multi-hop routing with power control in wireless sensor networks. *The Journal of China Universities of Posts and Telecommunications*, 14(1):49–57, 2007.

[114] Suneet Kumar Gupta, Pratyay Kuila, and Prasanta K. Jana. Energy efficient multipath routing for wireless sensor networks: A genetic algorithm approach. In *Advances in Computing, Communications and Informatics (ICACCI), 2016 International Conference on*, pages 1735–1740. IEEE, 2016.

[115] Md Azharuddin, Pratyay Kuila, and Prasanta K. Jana. A distributed fault-tolerant clustering algorithm for wireless sensor networks. In *Advances in Computing, Communications and Informatics (ICACCI), 2013 International Conference on*, pages 997–1002. IEEE, 2013.

[116] Dawei Gong, Yuanyuan Yang, and Zhexi Pan. Energy-efficient clustering in lossy wireless sensor networks. *Journal of Parallel and Distributed Computing*, 73(9):1323–1336, 2013.

[117] Suneet Kumar Gupta, Pratyay Kuila, and Prasanta K. Jana. Energy efficient routing algorithm for wireless sensor networks: A distributed approach. In *Communication and Computing Systems: Proceedings of the International Conference on Communication and Computing Systems (ICCCS 2016), Gurgaon, India, 9-11 September, 2016*, pages 207–213. CRC Press, 2017.

[118] Suneet Kumar Gupta, Pratyay Kuila, and Prasanta K. Jana. Ga based energy efficient and balanced routing in *k*-connected wireless sensor networks. In *Proceedings of the First International Conference on Intelligent Computing and Communication*, pages 679–686. Springer, 2017.

[119] Mong-Fong Horng, Yi-Ting Chen, Shu-Chuan Chu, Jeng-Shyang Pan, Bin-Yih Liao, Jang-Pong Hsu, and Jia-Nan Lin. A genetic algorithm with elite mutation to optimize cruise area of mobile sinks in hierarchical wireless sensor networks. In *International Conference on Computational Collective Intelligence*, pages 402–412. Springer, 2012.

[120] Kok-Lim Alvin Yau, Nordin Ramli, Wahidah Hashim, and Hafizal Mohamad. Clustering algorithms for cognitive radio networks: A survey. *Journal of Network and Computer Applications*, 45:79–95, 2014.

[121] Fan Ye, Haiyun Luo, Jerry Cheng, Songwu Lu, and Lixia Zhang. A two-tier data dissemination model for large-scale wireless sensor networks. In *Proceedings of the 8th annual international conference on Mobile computing and networking*, pages 148–159. ACM, 2002.

[122] Guangqian Xie and Feng Pan. Cluster-based routing for the mobile sink in wireless sensor networks with obstacles. *IEEE Access*, 4:2019–2028, 2016.

[123] Jia-Liang Lu and Fabrice Valois. On the data dissemination in wsns. In *Third IEEE International Conference on Wireless and Mobile Computing, Networking and Communications (WiMob 2007)*, pages 58–58. IEEE, 2007.

[124] Babar Nazir and Halabi Hasbullah. Mobile sink based routing protocol (msrp) for prolonging network lifetime in clustered wireless sensor network. In *Computer Applications and Industrial Electronics (ICCAIE), 2010 International Conference on*, pages 624–629. IEEE, 2010.

[125] Sylvia Ratnasamy, Brad Karp, Li Yin, Fang Yu, Deborah Estrin, Ramesh Govindan, and Scott Shenker. Ght: a geographic hash table for data-centric storage. In *Proceedings of the 1st ACM international workshop on Wireless sensor networks and applications*, pages 78–87. ACM, 2002.

[126] Elyes Ben Hamida and Guillaume Chelius. A line-based data dissemination protocol for wireless sensor networks with mobile sink. In *2008 IEEE International Conference on Communications*, pages 2201–2205. IEEE, 2008.

[127] Charalampos Konstantopoulos, Grammati Pantziou, Damianos Gavalas, Aristides Mpitziopoulos, and Basilis Mamalis. A rendezvous-based approach enabling energy-efficient sensory data collection with mobile sinks. *IEEE Transactions on parallel and distributed systems*, 23(5):809–817, 2012.

[128] Thomas H. Cormen, Charles E. Leiserson, Ronald L. Rivest, and Clifford Stein. *Introduction to Algorithms*. The MIT Press, 2009.

[129] Jamal N. Al-Karaki, Raza Ul-Mustafa, and Ahmed E. Kamal. Data aggregation and routing in wireless sensor networks: Optimal and heuristic algorithms. *Computer Networks*, 53(7):945–960, 2009.

[130] Robert J. Fowler, Michael S. Paterson, and Steven L. Tanimoto. Optimal packing and covering in the plane are np-complete. *Information Processing Letters*, 12(3):133–137, 1981.

[131] Rajgopal Kannan, Sudipta Sarangi, S Sitharama Iyengar, and Lydia Ray. Sensor-centric quality of routing in sensor networks. In *INFOCOM 2003. Twenty-Second Annual Joint Conference of the IEEE Computer and Communications. IEEE Societies*, volume 1, pages 692–701. IEEE, 2003.

[132] Mark Yarvis, Nandakishore Kushalnagar, Harkirat Singh, Anand Rangarajan, York Liu, and Suresh Singh. Exploiting heterogeneity in sensor networks. In *INFOCOM 2005. 24th Annual Joint Conference of the IEEE Computer and Communications Societies. Proceedings IEEE*, volume 2, pages 878–890. IEEE, 2005.

[133] Ataul Bari. *Energy aware design strategies for heterogeneous sensor networks*. PhD thesis, University of Windsor, Windsor, Ontario, Canada, 2011.

[134] Jianlin Mao, Zhiming Wu, and Xing Wu. A TDMA scheduling scheme for many-to-one communications in wireless sensor networks. *Computer Communications*, 30(4):863–872, 2007.

[135] Paolo Baronti, Prashant Pillai, Vince WC Chook, Stefano Chessa, Alberto Gotta, and Y Fun Hu. Wireless sensor networks: A survey on the state of the art and the 802.15. 4 and ZigBee standards. *Computer Communications*, 30(7):1655–1695, 2007.

[136] Pratyay Kuila and Prasanta K. Jana. Approximation schemes for load balanced clustering in wireless sensor networks. *Journal of Supercomputing*, 68:87–105, 2014.

[137] Pratyay Kuila and Prasanta K. Jana. Improved load balanced clustering algorithm for wireless sensor networks. In *International Conference Advanced Computing, Networking and Security (ADCONS 2011), LNCS (Springer)*, volume 7135, pages 399–404. Springer, 2011.

[138] Suneet Kumar Gupta, Pratyay Kuila, and Prasanta K. Jana. Genetic algorithm approach for k-coverage and m-connected node placement in target based wireless sensor networks. *Computers & Electrical Engineering*, 56:544–556, 2016.

[139] Suneet Kumar Gupta, Pratyay Kuila, and Prasanta K. Jana. Genetic algorithm for k-connected relay node placement in wireless sensor networks. In *Proceedings of the Second International Conference on Computer and Communication Technologies*, pages 721–729. Springer, 2015.

[140] Pratyay Kuila and Prasanat K. Jana. Energy efficient load-balanced clustering algorithm for wireless sensor networks. *Procedia Technology*, 6:771–777, 2012.

[141] Pratyay Kuila and Prasanta K. Jana. Heap and parameter-based load balanced clustering algorithms for wireless sensor networks. *International Journal of Communication Networks and Distributed Systems*, 14(4):413–432, 2015.

[142] Ilhem BoussaïD, Julien Lepagnot, and Patrick Siarry. A survey on optimization metaheuristics. *Information Sciences*, 237:82–117, 2013.

[143] Thitipong Navalertporn and Nitin V. Afzulpurkar. Optimization of tile manufacturing process using particle swarm optimization. *Swarm and Evolutionary Computation*, 1(2):97–109, 2011.

[144] Wei Hong Lim and Nor Ashidi Mat Isa. Particle swarm optimization with increasing topology connectivity. *Engineering Applications of Artificial Intelligence*, 27:80–102, 2014.

[145] Wi Sung Yoo, Hak-Ju Lee, Dong-In Kim, Kyung-In Kang, and Hunhee Cho. Genetic algorithm-based steel erection planning model for a construction automation system. *Automation in Construction*, 24:30–39, 2012.

[146] Issam Mazhoud, Khaled Hadj-Hamou, Jean Bigeon, and Patrice Joyeux. Particle swarm optimization for solving engineering problems: A new constraint-handling mechanism. *Engineering Applications of Artificial Intelligence*, 26(4):1263–1273, 2013.

[147] David E. Goldberg. *Genetic algorithms*. Pearson Education India, 2006.

[148] David E. Golberg. Genetic algorithms in search, optimization, and machine learning. *Addion Wesley*, 1989, 1989.

[149] Buthainah Fahran Al-Dulaimi and Hamza A Ali. Enhanced traveling salesman problem solving by genetic algorithm technique (tspga). *Proceedings of World Academy of Science: Engineering & Technology*, 40, 2008.

[150] Zakir H. Ahmed. Genetic algorithm for the traveling salesman problem using sequential constructive crossover operator. *International Journal of Biometrics & Bioinformatics (IJBB)*, 3(6):96, 2010.

[151] Jia-qing Zhao, Ling Wang, Pan Zeng, and Wen-hui Fan. An effective hybrid genetic algorithm with flexible allowance technique for constrained engineering design optimization. *Expert Systems with Applications*, 39(5):6041–6051, 2012.

[152] Adrian Dietz, Catherine Azzaro-Pantel, Luc Pibouleau, and Serge Domenech. Strategies for multiobjective genetic algorithm development: Application to optimal batch plant design in process systems engineering. *Computers & Industrial Engineering*, 54(3):539–569, 2008.

[153] Liu Guoshuai, Yang Kan, Zhou Ran, and Zheng Jiao. Application of genetic algorithm and annealing genetic algorithm in short-term optimal operation and economical operation of three gorges cascade. *Procedia Engineering*, 28:81–84, 2012.

[154] Saber M. Elsayed, Ruhul A. Sarker, and Daryl L. Essam. A new genetic algorithm for solving optimization problems. *Engineering Applications of Artificial Intelligence*, 27:57–69, 2014.

[155] Yong Wang. The hybrid genetic algorithm with two local optimization strategies for traveling salesman problem. *Computers & Industrial Engineering*, 70:124–133, 2014.

[156] Swagatam Das and Ponnuthurai Nagaratnam Suganthan. Differential evolution: A survey of the state-of-the-art. *Evolutionary Computation, IEEE Transactions on*, 15(1):4–31, 2011.

[157] Rainer Storn and Kenneth Price. Differential evolution–a simple and efficient heuristic for global optimization over continuous spaces. *Journal of Global Optimization*, 11(4):341–359, 1997.

[158] Zhihua Cai, Wenyin Gong, Charles X. Ling, and Harry Zhang. A clustering-based differential evolution for global optimization. *Applied Soft Computing*, 11(1):1363–1379, 2011.

[159] James Kennedy. Particle swarm optimization. In *Encyclopedia of machine learning*, pages 760–766. Springer, 2011.

[160] N.C. Sahoo, S. Ganguly, and D. Das. Simple heuristics-based selection of guides for multi-objective PSO with an application to electrical distribution system planning. *Engineering Applications of Artificial Intelligence*, 24(4):567–585, 2011.

[161] Abdullah Konak, David W. Coit, and Alice E. Smith. Multi-objective optimization using genetic algorithms: A tutorial. *Reliability Engineering & System Safety*, 91(9):992–1007, 2006.

[162] Daniel Bratton and James Kennedy. Defining a standard for particle swarm optimization. In *Swarm Intelligence Symposium, 2007. SIS 2007. IEEE*, pages 120–127. IEEE, 2007.

[163] Shao-Shan Chiang, Chih-Hung Huang, and Kuang-Chiung Chang. A minimum hop routing protocol for home security systems using wireless sensor networks. *Consumer Electronics, IEEE Transactions on*, 53(4):1483–1489, 2007.

[164] Jason Hill, Robert Szewczyk, Alec Woo, Seth Hollar, David Culler, and Kristofer Pister. System architecture directions for networked sensors. In *Proceedings of the ninth international conference on architectural support for programming languages and operating systems (ASPLOS '00)*, volume 34, pages 93–104. ACM, 2000.

[165] Bencan Gong, Layuan Li, Shaorong Wang, and Xuejun Zhou. Multihop routing protocol with unequal clustering for wireless sensor networks. In *Computing, Communication, Control, and Management, 2008. CCCM'08. ISECS International Colloquium on*, volume 2, pages 552–556. IEEE, 2008.

[166] Guihai Chen, Chengfa Li, Mao Ye, and Jie Wu. An unequal cluster-based routing protocol in wireless sensor networks. *Wireless Networks*, 15(2):193–207, 2009.

[167] Sungju Lee, Jangsoo Lee, Hongjoong Sin, Seunghwan Yoo, Sanghyuck Lee, Jaesik Lee, Yongjun Lee, and Sungchun Kim. An energy-efficient distributed unequal clustering protocol for wireless sensor networks. *Proceedings of World Academy of Science: Engineering & Technology*, 48, 2008.

[168] IEEE, Standard for Information Technology - Telecommunications and Information Exchange Between Systems - Local and Metropolitan Area Networks - Specific Requirements Part 15.4: Wireless Medium Access Control (MAC) and Physical Layer (PHY) Specifications for Low-Rate Wireless Personal Area Networks (WPANs). 2006.

[169] Pratyay Kuila and Prasanat K. Jana. An energy balanced distributed clustering and routing algorithm for wireless sensor networks. In *2nd IEEE International Conference on Parallel, Distributed and Grid Computing, 2012*, pages 220–225. IEEE, 2012.

[170] Md Azharuddin, Pratyay Kuila, and Prasanta K. Jana. Energy efficient fault tolerant clustering and routing algorithms for wireless sensor networks. *Computers & Electrical Engineering*, 41:177–190, 2015.

[171] Jiuqiang Xu, Wei Liu, Fenggao Lang, Yuanyuan Zhang, Chenglong Wang, et al. Distance measurement model based on RSSI in WSN. *Wireless Sensor Network*, 2(8):606–611, 2010.

[172] Indrajit Banerjee, Prasenjit Chanak, Hafizur Rahaman, and Tuhina Samanta. Effective fault detection and routing scheme for wireless sensor networks. *Computers & Electrical Engineering*, 40(2):291–306, 2014.

[173] Deepika Singh, Pratyay Kuila, and Prasanta K. Jana. A distributed energy efficient and energy balanced routing algorithm for wireless sensor networks. In *3rd International Conference on Advances in Computing, Communications and Informatics (ICACCI-2014), IEEE*, pages 1657–1663. IEEE, 2014.

[174] Srikanth Jannu and Prasanta K. Jana. A grid based clustering and routing algorithm for solving hot spot problem in wireless sensor networks. *Wireless Networks*, 22(6):1901–1916, 2016.

[175] Nabil Sabora, Mohammed Abo-Zahhad, Shigenobu Sasaki, and Sabah M. Ahmed. An unequal multi-hop balanced immune clustering protocol for wireless sensor networks. *Applied Soft Computing*, 43:372–389, 2016.

[176] Saim Ghafoor, Mubashir Husain Rehmani, Sunghyun Cho, and Sung-Han Park. An efficient trajectory design for mobile sink in a wireless sensor network. *Computers & Electrical Engineering*, 40(7):2089–2100, 2014.

[177] Can Tunca, Sinan Isik, M. Yunus Donmez, and Cem Ersoy. Distributed mobile sink routing for wireless sensor networks: a survey. *IEEE Communications Surveys & Tutorials*, 16(2):877–897, 2014.

[178] Gabriel T. Sibley, Mohammad H. Rahimi, and Gaurav S. Sukhatme. Robomote: A tiny mobile robot platform for large-scale ad-hoc sensor networks. In *Robotics and Automation, 2002. Proceedings. ICRA'02. IEEE International Conference on*, volume 2, pages 1143–1148. IEEE, 2002.

[179] Scott D. Hanford, Lyle N. Long, and Joseph F. Horn. A small semi-autonomous rotary-wing unmanned air vehicle (uav). *AIAA2005-7077*, 2005.

[180] Rahul C. Shah, Sumit Roy, Sushant Jain, and Waylon Brunette. Data mules: Modeling and analysis of a three-tier architecture for sparse sensor networks. *Ad Hoc Networks*, 1(2):215–233, 2003.

[181] Shashidhar Rao Gandham, Milind Dawande, Ravi Prakash, and Subbarayan Venkatesan. Energy efficient schemes for wireless sensor networks with multiple mobile base stations. In *Global telecommunications conference, 2003. GLOBECOM'03. IEEE*, volume 1, pages 377–381. IEEE, 2003.

[182] Guoliang Xing, Tian Wang, Zhihui Xie, and Weijia Jia. Rendezvous planning in wireless sensor networks with mobile elements. *IEEE Transactions on Mobile Computing*, 7(12):1430–1443, 2008.

[183] Hai Zhou, Narendra Shenoy, and William Nicholls. Efficient minimum spanning tree construction without Delaunay triangulation. In *Proceedings of the 2001 Asia and South Pacific Design Automation Conference*, pages 192–197. ACM, 2001.

Index